TRANSACTIONS

OF THE

AMERICAN PHILOSOPHICAL SOCIETY

HELD AT PHILADELPHIA
FOR PROMOTING USEFUL KNOWLEDGE

NEW SERIES—VOLUME 37, PART 1

POSTGLACIAL FOREST SUCCESSION, CLIMATE, AND CHRONOLOGY IN THE PACIFIC NORTHWEST

HENRY P. HANSEN

Associate Professor of Botany
Oregon State College

THE AMERICAN PHILOSOPHICAL SOCIETY
INDEPENDENCE SQUARE
PHILADELPHIA 6

1947

COMMITTEE ON PUBLICATIONS

LANCASTER PRESS, INC., LANCASTER, PA.

FIG. 1. View of Newman Lake near Spokane, Washington. This lake was ponded in a short tributary of the Spokane River by aggrading of the main stream with glacial outwash, damming the mouth of the tributary. Adjacent slopes are forested largely with western yellow pine.

FIG. 2. Patchwise stands of sub-alpine species in southern Washington Cascades.

POSTGLACIAL FOREST SUCCESSION, CLIMATE, AND CHRONOLOGY IN THE PACIFIC NORTHWEST [1]

Henry P. Hansen

CONTENTS

[1] Approved by the Monographs Publication Committee, Oregon State College, as research paper No. 105, School of Science, Department of Botany.

1

ACKNOWLEDGMENTS

The writer is grateful to many persons for aid and assistance during the several years this work has been in progress. Especial thanks are due Dr. Ira S. Allison, Oregon State College, for reading and criticizing the entire manuscript during its course of preparation, and Dr. Ernst Antevs, Globe, Arizona, for reading and criticizing the chapters on chronology and climate, and for additional data on both European and North American late-glacial and postglacial chronology. I thank Dr. G. B. Rigg, University of Washington for assistance and suggestions in the early stage; Mr. John Broadbent, U. S. National Park Service, Port Angeles, Washington, for collection of fresh pollen from various conifers; Mr. T. T. Munger and Mr. S. N. Wykcoff, Pacific Northwest Forest and Range Experiment Station, Portland, Oregon, for use of certain maps; Mr. Karl Janouch, Supervisor, Rogue River National Forest, for arrangements to get fresh pollen of certain conifers; and Dr. W. M. Atwood, Oregon State College, for assistance in photographic work.

I wish to express my sincere thanks to Dr. P. B. Sears, Oberlin College, for critically reading the manuscript and offering valuable suggestions; to Dr. M. E. Peck, Willamette University, for reading the chapters on present vegetation and vegetation history and for his important criticisms and suggestions; to Dr. L. S. Cressman, University of Oregon, for reading the manuscript and for his criticisms, particularly con-

cerning early man in Oregon; and to Dr. F. A. Gilfillan, Dean of Science, Oregon State College, for certain courtesies and encouragement in pursuing the problem.

I thank many persons who have helped in a multitude of minor ways, such as assistance in obtaining sedimentary columns, preparing peat samples, typing, drawing, etc.

To my wife, Helen Rivedal Hansen, for assistance in the field, reading the manuscript, and for constant inspiration, I express my deep sense of appreciation.

Grants-in-aid from the General Research Council of the Oregon State System of High Education made possible collection of much of the field data, purchase of certain equipment, and clerical assistance, for which I am grateful.

An appointment to a fellowship of the John Simon Guggenheim Memorial Foundation for the year of September 1943 to September 1944 made it possible to spend the entire year in collecting additional field data and sedimentary columns, completing analyses of sedimentary columns, making an intensive study of the data, reading of much literature, and completing most of the manuscript. This would have been impossible without devoting one's entire time. I wish to express my sincere thanks and appreciation to Mr. Henry Allen Moe, Secretary General, and to the Committee of Selection for this honor and opportunity.

INTRODUCTION

During the Pleistocene a vast area of temperate North America was covered several times by great ice sheets that profoundly altered the topography and caused displacement and readjustment of the vegetation. The expansion of the glaciers caused the forests to retreat, and upon recession of the ice the forests reinvaded the areas left denuded in the wake of the ice. The forests in the vanguard were composed of species that could withstand the rigorous environment which must have existed near the ice front. These conditions were not only climatic, but also edaphic and physiographic, resulting from geomorphic instability incident to the melting ice. As the physiographic conditions were stabilized and the soil and climate ameliorated by the pioneer vegetation, other species followed and expanded at the expense of those that had modified the environment. These in turn were wholly or partially replaced by others, and a continual process of coaction between the plants and reaction between the plants and the physical environment culminated in the forests that were present when the white man arrived on the scene. In the Pacific Northwest only northern Washington and Idaho and the Cascade Mountain range were glaciated.

In the wake of the retreating ice and its geomorphic influence many lakes and ponds were formed. Beyond the limits of glaciation many lakes also were formed owing to geomorphic and climatic cycles initiated by the influence of advancing and retreating glaciers. Today many of these lakes have become filled with sediments, largely organic, which abound with pollen grains from adjacent forests. This pollen settled annually in the lake and later on the bog surface and has been preserved to the present. These sediments, therefore, hold a record of the general, major changes in the composition of these forests, beginning with those which followed in the wake of the retreating ice and continuing to the present. In the non-forested areas a record of the lesser vegetation is interred. Samples of these sediments taken at suitable intervals from bottom to top, and analyzed for their pollen content, reveal that as postglacial time progressed, there were gradual and occasionally abrupt changes in the pollen proportions of the several species from level to level. Some species predominantly represented at the bottom horizon may be entirely absent in the higher levels, while other species record a reverse trend. It is assumed that these changes in pollen proportions denote a change in the composition of the forests within range of pollen dispersal to the site of the sediments. Alteration in forest composition sustained for an appreciable period of time and consistently portrayed in many sedimentary columns from a region denotes environmental changes. Some of these changes were undoubtedly due to climatic cycles which have been regional and are systematically recorded in many sedimentary columns from the entire Pacific Northwest. Other trends reflect normal plant succession which modified the environment and paved the way for successive stages of vegetation. Yet other fluctuations in the pollen proportions from level to level indicate local changes in the environment caused by fire, disease, and volcanic activity.

The most important postglacial volcanic activity, and that which exercised the greatest influence on postglacial vegetational succession, was the eruption of Mount Mazama in the Cascades of southern Oregon (Williams, 1942). This great eruption resulted in the formation of the caldera holding Crater Lake, and spread a pumice mantle six inches or more deep for a hundred miles north and east, covering an area at least 5,000 square miles in extent. This volcanic activity is dated at about 10,000 years ago, as will be discussed later. Other volcanic activity in the central Cascades of Oregon has had a local effect upon the vegetation. The occurrence of a stratum of volcanic ash in at least thirty postglacial sedimentary columns located in Washington, Idaho, and British Columbia records an eruption that dispersed ash over a wide area. The source of this ash seems to be located in north central Washington, probably Glacier Peak where there is an abundance of pumice overlying the youngest glacial drift (Waters, 1939). In a personal communication, Dr. A. C. Waters states that this pumice grades into a thin layer of white ash eastward, and that he has traced the ash all the way eastward to Spokane. This is consistent with the fact that the thickest strata of ash occur in sediments lying to the east and southeast of Glacier Peak, attaining their greatest thickness directly east and extending into northern Idaho. In the Puget Lowland, west of the Cascades and Glacier Peak, the ash layer is thinner and in some bogs occurs only as sparsely scattered glass fragments. This suggests that during the eruption strong westerly winds carried the bulk of the ejecta eastward. The apparent absence of the ash in bogs along the coast and in southwestern Washington further supports this view. Although there are several extinct volcanic cones in the Washington Cascades, the presence of a single layer in so many widely scattered columns, the similarity of the ash, and its apparent center of distribution all tend to indicate that the ash is from a common and simultaneous source. Unlike that of Mount Mazama, the eruption of Glacier Peak had little or no effect upon the vegetation, except perhaps on the immediate slopes. The greatest significance of the Washington volcanic ash is its tremendous value as a chronological indicator. Its occurrence in so many columns provides a common time marker and serves to correlate both chronology and forest succession over the entire region. It is dated at about 6,000 years, as will be discussed later. In the Oregon Cascades and in the northern Great Basin of south central Oregon, pumice strata from several sources in sedimentary columns serve as valuable time markers in this region also.

The ages of the sedimentary columns that rest upon glacial drift or its chronological equivalent perhaps range from 15,000 to 25,000 years, depending upon the location of their sites in relation to glacial movements and geomorphic cycles brought about by glaciations. Although there are several phytogeographic and climatic provinces in the Pacific Northwest, pollen analytical data from about seventy sedimentary columns indicate that the postglacial climatic trends have been much the same over the entire region, although varying in degree. Upon the basis of the forest succession as expressed in the pollen profiles, the postglacial climatic trend has been divided into four main stages. The first stage persisted until 15,000 years ago and was cooler and moister than the present. Some of the sedimentary columns may represent 5,000 to 10,000 years of this initial stage while others may hardly have had their inception by its termination. The second stage was one of increased warming and drying until about 8,000 years ago. The third stage was one of maximum warmth and dryness and endured until about 4,000 years ago, while the final stage has been cooler and more humid than the preceding.

The period of maximum warmth and dryness between 8,000 and 4,000 years ago was evidently general throughout the north temperate zone, as is suggested by pollen profiles from northern Europe, England, eastern North America, and the Great Lakes region, and by evidence from other sources. The warm, dry stage is not manifested near the Pacific Ocean and only to a slight degree in the Puget Lowland. In the Willamette Valley of western Oregon it is more strongly pronounced, while it is best expressed in profiles from the yellow pine forests and timberless zones of eastern Washington and Oregon.

The position of human remains, artifacts, and other evidence of human occupation, associated with fossils of contemporary fauna, in relation to the pumice, tells us that early man lived in southern Oregon prior to the eruption of Mount Mazama, and that he survived the warm, dry period in the northern Great Basin.

In the course of his studies of the Tamarack bogs of the Driftless Area of Wisconsin (Hansen, 1933), the author became interested in the bogs of the Pacific Northwest through reading of the bog literature. Pollen analysis of peat bogs and the interpretation of postglacial vegetation had just begun in the United States and Canada, and had aroused the interest of workers as far west as Wisconsin and Iowa. During the autumn of 1935 the author made his first observations of peat bogs in the Pacific Northwest of America. He was very fortunate in having Dr. G. B. Rigg, Professor of Botany, University of Washington, Seattle, show him the location of many bogs in the Puget Sound region. Professor Rigg has been the pioneer investigator in bog studies in the Pacific Northwest, and the writer greatly benefited by the wealth of information that was made available by him. Professor Rigg's first studies on Pacific Northwest bogs go back to 1913, and since then he has published many articles on bog floristics, peat types, organic sedimentation, and peat profiles from this region. He has also studied bogs in many parts of North America, and his comparisons of Sphagnum bogs in various regions have been a valuable contribution to the knowledge of bog characteristics and development (Rigg, 1940a, 1940b).

The first studies undertaken by the writer were pollen analyses of two bogs located in the Puget Sound region (Hansen, 1938). Previously, Osvald (1936) made an investigation of stratigraphy and pollen flora of some bogs in this region, but the writer has been unable to obtain the results of this study. The only known subsequent pollen studies are those made by the author. Before pollen analysis was begun, the collection and study of fresh pollen from the forest trees of the Pacific Northwest were necessary. Pollen of other plants that might be present in peat was also collected. This in itself was a task that involved considerable time and travel, as there had been little published on forest tree pollen of the Pacific Northwest. In this work "Pollen Grains" by Wodehouse (1935) was of invaluable assistance. The author had had a little previous experience in pollen analyses of peat bogs in Wisconsin (Hansen, 1937), so the problems of pollen analysis were not entirely new. This experience, although concerned with entirely different species, enabled the writer to begin the actual pollen analysis of sedimentary columns with less preliminary work than otherwise would have been required. Another factor in the problem of pollen analysis in a new region was the lack of knowledge of the forest tree characteristics and ecological requirements. This knowledge was necessary in order to interpret the meaning of the pollen profiles and to define them in terms of forest succession and climate. As more knowledge has been obtained through field experience and reading of available literature, more definite and tenable interpretations have been applied to the pollen profiles. This has in a few cases necessitated the revision of earlier concepts, and has resulted in certain inconsistencies. These have not been of serious nature, and will be corrected in the course of this paper.

As is often the case in the initial work in a broad field of scientific endeavor, the early pollen analyses of peat bogs were carried out without full realization of the scope and other potentialities of the problem in the Pacific Northwest. Certain data were overlooked in these early studies that seemed of little significance in terms of only a few profiles, but which assume a more important status when applied to the results and correlations of many sedimentary columns over a wide area of diverse climatic and vegetation characteristics. Additional data have since been collected on other visits to the sites of the bogs.

Most of the work in the reconstruction of postglacial vegetation by pollen analysis of sedimentary columns has been done in Europe. In North America most of the studies in this field have been made in the East and the Middle-west, largely perhaps, because of the abundance and accessibility of peat deposits. Pollen analysis in the Pacific Northwest, however, has progressed at a much slower pace, owing to the absence of workers, the comparative scarcity of pollen-bearing sediments, the vast areas involved, and in many cases the inaccessibility of the sites of organic sedimentation. In the collection of about seventy sedimentary columns scattered throughout the Pacific Northwest, and in this study comprising the states of Oregon, Washington, northern Idaho and southwestern British Columbia, the author has travelled some 16,000 miles by automobile, not to mention many miles on foot. Some of this travel has been in vain because of failure to reach a bog owing to mountainous terrain, or to the unsuitability of the sedimentary column for pollen analysis. In spite of these difficulties an attempt has been made to obtain a series of profiles that perhaps hold a fairly representative picture of the postglacial vegetation of the Pacific Northwest. Sedimentary columns have been obtained from most of the natural vegetation areas that are more or less homogeneous with respect to physiography and physiographic history, climate, and flora. These aims have been better attained than at first thought possible, and, in spite of the location of bogs largely in areas of peculiar physiographic history, sedimentary columns have been obtained from areas hardly suspected of supporting hydrarch succession.

The results and interpretations of pollen analyses of single sedimentary columns have been published by the author in individual papers. He has been loath to do this, feeling that several profiles from a given natural area offer more reliable data for interpretation of postglacial forest succession. Fortunately, further studies of profiles in the same or similar phytogeographic provinces have borne out the findings of the initial analyses. The absence from this region of other workers in the field of pollen analysis is unfortunate because one would feel much surer about his own results if they could be either refuted or substantiated by the unprejudiced work of others. A lack of knowledge of the ecological requirements of the significant species and groups of species has also been a handicap in the interpretation of the forest succession from the pollen profiles. The successional status of each species as well as groups of species should be fairly well realized before attempts are made to define the pollen profiles in terms of vegetation succession and climatic trends. The occurrence of

certain species in several phytogeographic provinces with different associates and under diverse environment tends to complicate the interpretation of postglacial vegetation upon the basis of pollen analysis. It is believed that this complication is of greater magnitude in the Pacific Northwest than in the Middle-west or Eastern United States. The postglacial forests in the latter region have evidently been more homogeneously distributed over a wider area because of the more homogeneous physiographic and climatic conditions. In the Pacific Northwest the great relief and varying topography, the vastly different physiographic history of the several physiographic provinces, and the wide range of climate, have resulted in the development of several different forest complexes.

Although sedimentary columns have not been obtained from all the climatic and phytogeographic provinces in the Pacific Northwest, the seventy sections furnishing the pollen analytical data of this study are sufficiently widely scattered to be representative of most of this vast region. The picture of postglacial vegetational succession and climatic trends is so consistently portrayed that the author is confident that the results and conclusions are reliable. Pollen analysis of additional sections may provide data for local trends and more refinement of the interpretations at hand, but would probably not change the overall picture of the major, general vegetational and climatic cycles.

HISTORY AND THEORY OF POLLEN ANALYSIS

The most recently interred records of prehistoric and historic biota, namely those of the post-Pleistocene, have been the last to receive the consideration of paleontologists. Not until well into the twentieth century have intensive and extensive studies been made of the microfossils that are preserved in the many kinds of postglacial sediments. These sediments for the most part are obvious, rather easily available, of abundant and widespread distribution, and perhaps hold a more detailed, short-period record of the fluctuations and adjustments of adjacent life, particularly those of forest trees, than most older fossil-bearing strata. Whereas modern pollen analysis with its applications and correlations made its debut in 1916, at Oslo, Norway, in the presentation of a paper on fossil pollen analysis by L. von Post (Erdtman, 1943), this field of scientific endeavor had its earliest known inception in 1885 by a Swiss geologist, J. Früh. During the interim from 1885 until 1916 the development of interest in pollen analysis and the realization of the value and potentialities of this paleontological method underwent slow incubation. Since 1916 a tremendous amount of work has been done in northern Europe, and pollen analysis has reached a mature stage involving the refined and detailed work of a science that has proved its value. Erdtman (1943) presents a detailed account of the history of the European work as it has progressed to the present. Another lapse of time, about ten years, took place before recognition of the significance and value of the method spanned the Atlantic Ocean and was applied to the post-Pleistocene sediments of North America. Early workers include Auer (1927, 1930), Draper (1928), Sears (1930), and Voss (1931). Since these earlier investigations the more prolific workers have been Sears, Wilson, Potzger, Voss, Cain, and Deevey, while others have worked out one or a few profiles under these investigators for theses in their academic requirements.

The theories and principles of pollen analysis are essentially as follows:

1. Each year the vegetation of a given area produces pollen, and most of the dominant forest trees of the north temperate zone produce windborne pollen.

2. The pollen of the anemophilous species is disseminated to some distance, depending upon the efficiency of the adaptation for dispersal, the weather during the period of anthesis, and the air conditions at the time of dissemination.

3. The pollen eventually settles upon the substratum and if it falls upon the surface of a pond it may settle to the bottom and become incorporated in the accumulating sediments. If it falls upon a bog surface it may become preserved in the accumulating peat. If the conditions are favorable and remain so, the pollen may be preserved indefinitely. Year after year, decade after decade, century after century, and millennium after millennium, this annual process takes place, and as the sediments are built up a record of the adjacent vegetation is interred.

4. If appreciable changes in the composition of the adjacent vegetation occur, a corresponding adjustment takes place in the proportions of the pollen grains of the several species preserved in the sediments.

5. Thus, as the composition of the dominant vegetation changes to a substantial degree over a period of time, whatever may be the cause, these changes are recorded and expressed in fluctuations in the pollen profile of each species.

6. The pollen grain characteristics of most genera of plants are sufficiently distinct to recognize them. Also the pollen grains of many species can be distinguished.

7. By obtaining samples of pollen-bearing sediments at intervals from bottom to top in a sedimentary bed, and preparing the sediments for microscopic analysis, the fossil pollen grains may be identified and counted.

8. The percentage of the total forest tree pollen for each species is determined and depicted in a pollen diagram composed of the pollen profiles of those species that are indicators of forest succession and climatic trends.

9. The forest succession during the period represented by the section of the sedimentary bed is then interpreted from the fluctuations in the pollen proportions from level to level. In turn, the climatic or other environmental changes are deduced from the interpreted forest succession.

The fundamental principles of pollen analysis as listed above are at first glance simple and obvious enough. Further consideration, however, reveals a complex of factors, many of them immeasurable and some even intangible, that influences the picture of forest succession and climate. The results and conclusions of pollen analysis, at best, can be interpreted in only general and relative terms. These interpretations, however, are much strengthened and supported by corroborative evidence from other sources, as well as by a remarkable consistency in the recorded trends in many sedimentary columns from diverse phytogeographic and climatic provinces throughout the Pacific Northwest. In such researches, as based upon the theories and principles listed above, it is obvious that certain assumptions must be made. Although these assumptions have considerable support in other evidence, and are entertained with confidence as to their reliability, they can hardly be held with absolute certainty.

In general it is assumed that the number of pollen grains of the several species at a given level is directly proportional to the abundance of those species that existed within range of pollen dispersal to the site of the sediments at that time. Although the pollen analyst interprets the vegetational succession upon this basis, he realizes that there are many factors that tend to invalidate it. It is a well known fact that all species do not produce pollen in proportion to their abundance and importance in the vegetation complex. In the Pacific Northwest, for instance, the writer knows that lodgepole pine produces more pollen than most of the other forest trees, and out of proportion to its abundance in the forest. On the other hand, grassland does not produce nearly so much pollen per unit of area as does a conifer forest, and thus is under-represented in the pollen diagrams. Under-representation may be further magnified by the fact that its pollen is not so well adapted for dispersal. A slight fluctuation in the number of grass pollen grains from level to level is more significant than one of similar degree of forest tree pollen, and must be interpreted accordingly.

This problem, however, may be solved to some extent by a comparison of the composition of the pollen rain with that of the forest in the same general area. Such studies have been carried out in Europe to some extent (Lüdi, 1937; Firbas, 1934; Erdtman, 1931), but only one such study has been made in this country (Carroll, 1943). The latter is concerned with the spruce-fir forest on Mount Collins in the Great Smoky Mountains. By pollen analysis of bryophytic polsters, it was found that the pollen proportions compared favorably with the density of spruce and fir of 10 inches d.b.h. or over, while birch was somewhat over-represented.

The reliability of pollen diagrams in depicting paleic forest successions may be further modified by the percentage of pollen grains produced that reach the site of sedimentation. This is dependent upon the relative efficiency of pollen dispersal, distance from the sediments, the relief of the adjacent terrain, the direction of the wind at the time of pollen shedding, and other factors. After the pollen reaches the bog, swamp, or lake, the relative degree of preservation also enters into the picture. Western red cedar, juniper, and other species of this group apparently have pollen that is not well preserved in organic sediments.

The transport of pollen by water into swamps that are seasonally inundated by flood water from melting snow causes over-representation of species that occur at great distances, perhaps beyond the limits of pollen dispersal by air. This factor probably enters into the picture of vegetational succession in the Channeled Scablands of eastern Washington, where streams that head in forested areas flood swamps that lie many miles beyond the forests. The pollen grains may be air-borne for a considerable distance, and then, settling in a stream, they may be carried well beyond their normal radius of distribution. The location of bogs in mountains also undoubtedly modifies forest tree representation in the pollen profiles. Two or three life zones may be in close proximity to the site of the sediments, and their positions at greater altitudes permit downward drifting of pollen onto the lake or bog. In lowland areas the homogeneity of the vegetation over a much wider area results in a more representative pollen rain. The proximity of the site of sedimentation to the boundary of a phytogeographic province in relation to the direction of the prevailing winds during anthesis also tends to result in misrepresentation. On the coast the location of the lodgepole pine forests windward to the bogs has favored over-representation of this species. Yet other physical and biological factors in the region and at the site of sedimentation may radically influence the accuracy of representation.

In spite of these factors that may cause misrepresentation of the adjacent vegetation in the pollen profiles, it would seem that the degree of error is somewhat reduced by the magnitude of the time and space involved. A small unit thickness of the sedimentary column constitutes many years of deposition. The variable record of the seasonal changes and local and short period fluctuations due to physical factors is somewhat evened. Furthermore, interpretation of the record must be in general terms, assigning small

fluctuations back and forth and from level to level to inherent peculiarities of the method.

The interpretation of climatic trends and other environmental changes from the indicated plant succession is one of the primary objectives of pollen analysis To do this, one must assume that species in the past reacted to environmental changes in much the same way as today. The historic observation of plant succession is limited with respect to the period of time involved. No one has observed succession for as long as a millennium, nor even for a few centuries, while historical accounts are meager. It is assumed that the composition of paleic forests as recorded in the sedimentary column reflects much the same environment as a similar composition does today. Minor local changes are recorded and are difficult to interpret. The significance of these local, short time variations is minimized, however, when many sedimentary columns in a homogeneous region are analyzed, presenting a picture of the major, long range cycles over a wide area. The minor, local changes are absorbed in the overall, major, long period cycles. The interpretations are further strengthened by analyses of many sections, involving different species, from several diverse climatic and phytogeographic provinces. This is true in this study, and the consistency of the indicated trends leaves little doubt in the writer's mind as to the reliability of the record. Pollen profiles from sagebrush areas of eastern Washington portray the same trends as those from the Willamette Valley; pollen profiles from the yellow pine forests of the Oregon Cascades and eastern Washington reflect the same cycles as those from the Puget Lowland, while the pollen records from the northern Great Basin of south central Oregon present a picture similar to that of northern Idaho.

Another consideration in pollen analysis that must be based to some extent upon assumption, however well supported, is chronology. While not absolutely essential, chronology adds considerable interest and value to the results and conclusions, and also serves to integrate the picture over the entire region. A correctly interpreted and applied chronology also serves better to correlate more events and trends over a much greater area. Fortunately there is a series of time markers more or less common to large areas of the Pacific Northwest. The most important is Pleistocene glaciation. Studies in Pleistocene and postglacial geomorphology indicate that the sedimentary columns providing the pollen analytical data for this study had their origins some time after the maximum of the last Wisconsin glaciation. Chronologic studies and correlations of the last glacial maxima in various parts of the world suggest that this occurred about 25,000 years ago (Antevs, 1945). Pollen-bearing sediments that rest upon drift from this glacial stage or its chronological equivalent are necessarily not more than 25,000 years old. How much younger they may be is hard to estimate because of local differences in ice positions, rate of ice retreat, persistence of dead ice, inundation, erosion, ice drainage, and other conditions determining the time of earliest sedimentation and migration of vegetation into deglaciated terrain. Obviously the columns vary considerably in age, even within a small area.

The age of the pollen-bearing sediments situated beyond the limits of Pleistocene glaciation is estimated as about the same as that of those that lie upon glacial drift, except those that can be dated by local conditions or events. The former are dated as postglacial because the geomorphic cycles responsible for the formation of the depressions were initiated and controlled by glaciation in other parts of the continent.

GEOLOGIC ORIGIN AND AGE OF THE PACIFIC NORTHWEST LAKE BASINS

GENERAL STATEMENT

Lakes and ponds in North America are distributed largely within the limits of Pleistocene glaciation, and their origins are related chiefly to glacial processes. In fact, the formation of most existing lakes may be traced back directly or indirectly to changes in topography wrought by glaciation. This is true even of many lakes that lie well beyond the boundaries of glaciation as well as those lying directly upon glacial deposits. The greatest number and concentration of bogs, swamps, and other sources of pollen-bearing sediments are located within the glaciated region, and their beginning generally dates back to the time when their sites were freed of ice. Thus, the largest number of mature peat profiles suitable for pollen analysis occur in the north central and northeastern parts of the United States where Pleistocene glaciation took place on a grand scale. As lakes, ponds, and other types of standing water provide the sites for the accumulation of sediments which serve as a basis for this study, it seems pertinent to discuss briefly the lake types and their geologic origin and setting in the Pacific Northwest. Of about sixty such lakes practically all owe their origin directly or indirectly to glaciation, including those that occur beyond the region covered by the continental ice sheets. About one-third fall in this latter category. The significance of such a discussion becomes more apparent when the variation in the genesis of lakes in this region is considered. The types of sedimentary basins and their manner of origin are also of value in chronological interpretation of the postglacial forest succession and climatic trends from the pollen profiles.

PUGET SOUND REGION

Only a small portion of the Pacific Northwest was subjected to Pleistocene continental and mountain glaciation (map 1). The total area covered by the mountain glaciers was probably greater than that overridden by the continental ice sheet. In the Puget Sound region, drift from at least two glacial stages has been noted (Willis, 1898; Bretz, 1913). The later stage is probably correlative with the maximum of the "Wisconsin" ice sheet (Antevs, 1929). The ice moved in a general southerly direction as a tongue of the Cordilleran ice cap. The glacial till and glaciofluvial deposits become progressively thinner southward, the southern limit of this mantle being about 15 miles south of Puget Sound. The morainic border follows roughly the divide which separates the Puget Sound drainage from the basin of the Chehalis River, but the Chehalis valley contains considerable outwash carried westward toward the Pacific Ocean.

The degree of glaciation was generally insufficient to modify greatly the preglacial topography or to provide as many favorable sites for organic sedimentation as in north central and northeastern United States, except, perhaps, in a few localities. In general, much of the topography is either young or mature, with comparatively little standing water available for hydrarch succession and resultant accumulation of pollen-bearing sediments. The greatest number and concentration of lakes, bogs, and swamps occur in the Puget Lowland of western Washington. This region was subjected to at least two stages of Pleistocene glaciation, and most of the lakes and ponds still extant or filled with organic sediments are genetically related to glacial topography. The deposition of a thick mantle of glacial till and glacial drift has favored the formation of several types of glacial lakes. These include (1) kettle lakes distributed most abundantly in outwash, both valley and plain, (2) morainal lakes, (3) lakes formed in drainage channels obstructed with glaciofluvial deposits, and (4) those formed in floodplain depressions in valleys occupied by glacial streams. Many of the small lakes have become filled with organic sediments, while others support all stages of hydrarch succession from submerged to climax seres. Peat deposits also occur on the beds of proglacial lakes, where shallow depressions have maintained a water table sufficiently high to support swamp vegetation. There are very few peat deposits in the Puget Sound region that have accumulated in stream channels, oxbow lakes, and other types of depressions formed by meandering or braided streams on aggraded floodplains.

NORTHERN WASHINGTON AND IDAHO

Northern Idaho and the northern part of Washington lying east of the Cascade Range were also subjected to Pleistocene glaciation (map 1). In this region the last glaciation is generally considered to be correlative with the last Wisconsin stage, the Mankato. The main north-south valleys carried lobes of the Cordilleran ice sheet, two of which, the Okanogan and Spokane lobes, encroached upon the Columbia

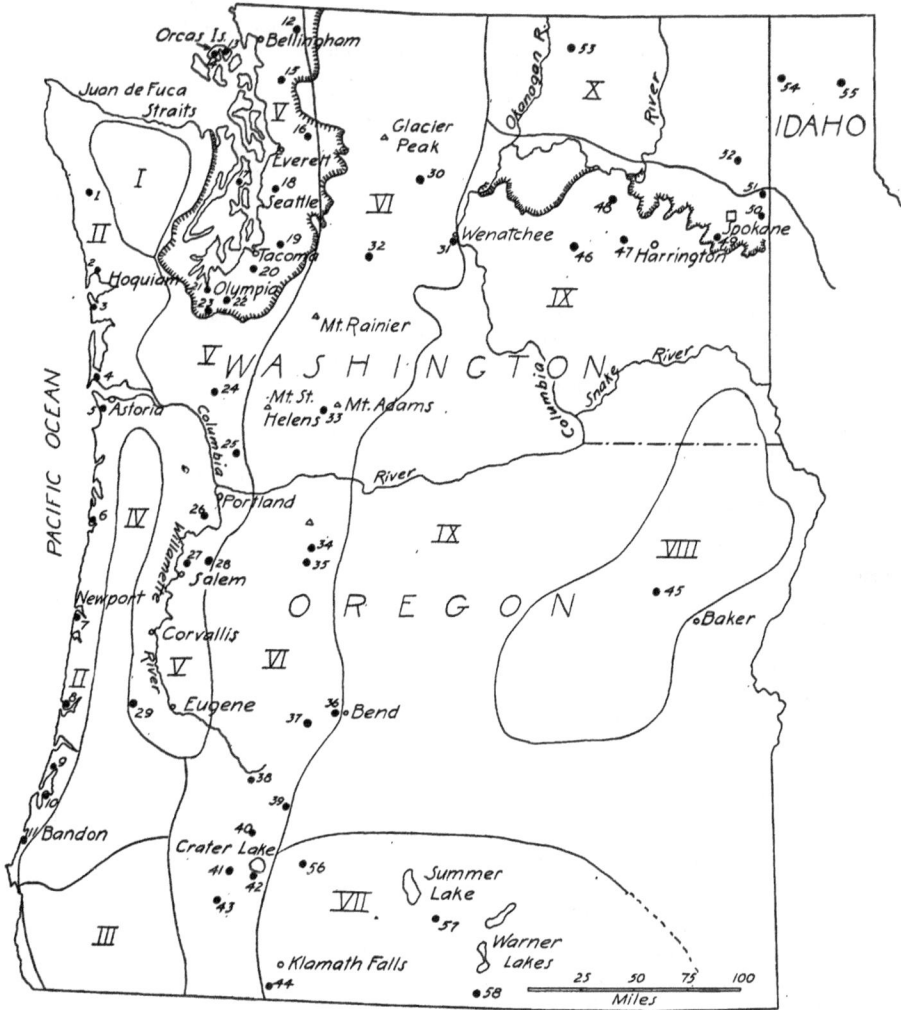

MAP 1. Oregon and Washington and northern Idaho, showing natural areas, continental glacier boundaries, and location of sites from which sedimentary columns were obtained. The natural areas approximate physiographic divisions, but they are delimited largely upon the basis of homogeneity of vegetation and climate, and to some extent by the geologic history. I. Olympic Mountains. II. Coastal Strip: 1, Forks; 2, Hoquiam; 3, Grayland; 4, Ilwaco; 5, Gearhart; 6, Sandlake; 7, Newport; 8, Woahink Lake; 9, Hauser; 10, Marshfield; 11, Bandon. III. Klamath-Siskiyou. IV. Oregon Coast Range. V. Puget-Willamette Lowland: 12, Bellingham; 13, Mt. Constitution; 14, Killebrew; 15, Sedro-Woolley; 16, Granite Falls; 17, Poulsbo; 18, Ronald; 19, Black Diamond; 20, Parkland; 21, Olympia; 22, Rainier; 23, Tenino; 24, Silver Lake; 25, Farger Lake; 26, Onion Flats; 27, Lake Labish; 28, Scotts Mills; 29, Noti. VI. Cascade Mountain Range: 30, Fish Lake (near Lake Wenatchee); 31, Wenatchee; 32, Lake Kaches; 33, Cayuse Meadows (Mt. Adams); 34, Clear Lake; 35, Clackamas Lake; 36, Bend (Tumalo Lake); 37, Mud Lake; 38, Willamette Pass; 39, Big Marsh; 40, Diamond Lake; 41, Rogue River; 42, Munson Valley; 43, Prospect. VII. Northern Great Basin: 44, Klamath Falls; 56, Klamath Marsh; 57, Chewaucan Marsh; 58, Warner Lake. VIII. Blue Mountains: 45, Anthony Lakes. IX. Columbia Basin: 46, Crab Lake; 47, Harrington; 48, Wilbur; 49, Cheney. X. Northeastern Washington and northern Idaho: 50, Liberty Lake; 51, Newman Lake; 52, Eloika Lake; 53, Bonaparte Lake; 54, Priest Lake; 55, Bonners Ferry. Hachured line represents glacial boundary.

Basin. Bretz (1923) believed that the Spokane lobe was pre-Wisconsin or early Wisconsin, whereas Flint (1937) holds that the ice margin was contemporaneous throughout its length, that its date was probably Wisconsin, and that it apparently was the last glaciation involved in this region. All the drift thus far examined by Flint in eastern Washington is said to date from the same glaciation although deposited at different times during that glaciation (Flint, 1937). It would then be considered as the equivalent of the Vashon stage in the Puget Lowland.

Northern Washington and northern Idaho, however, do not provide so many sites favorable for hydrarch succession as does the Puget Sound region. The more rugged topography was not sufficiently altered to effect ponding of water, nor was the glacial drift deposited in a manner suitable for the formation of large numbers of small lakes. The somewhat drier postglacial climate may also have been unfavorable for maximum organic sedimentation. Several large lakes, including Pend Oreille, Priest Lake, and Coeur d'Alene, are of glacial origin, but apparently few or no organic sedimentary columns suitable for pollen analysis have developed. The type of lake most prevalent in the region is that formed by aggrading of main stream valleys with glacial outwash, so as to blockade the mouths of the tributaries where they enter the main valleys, there ponding them (figs. 1, 43). These lakes are usually situated above the present level of the main valley floor and, having little or no outward drainage on the surface, provide favorable sites for organic sedimentation. These lakes occur in the tributaries of valleys occupied in part by the larger ice lobes of the last glaciation. The other chief type of lake of glacial origin is the kettle pond lying largely in valley outwash, while at least one recessional moraine lake was noted in the Okanogan Highlands. A few lakes and swamps have been formed on aggraded floodplains in oxbows, abandoned channels, between alluvial fans, and in other types of floodplain depressions, but none of the several observed has developed organic sediments suitable for pollen analysis. The most extensive of these are in the Kootenai River and Coeur d'Alene River valleys. In the Okanogan Highlands of northern Washington sites favorable for organic sedimentation seemed to be almost entirely absent.

CHANNELED SCABLANDS OF COLUMBIA BASIN

That part of the Columbia Basin bounded on the north by parts of the Columbia and Spokane Rivers and extending southwestward to the Snake River is spoken of as the Channeled Scablands. This term was first applied by Bretz (1923) to designate the northeastern part of the Columbia Basin that was severely eroded by meltwater from the receding glaciers. During glaciation the Columbia River was dammed by ice lobes, and marginal lakes along the ice front rose sufficiently to spill over the divide south of the Spokane and Columbia Rivers. The water developed great erosive power as it gained velocity flowing down the sloping plateau toward the Columbia and Snake Rivers. This area is characterized by a system of interlacing channels or coulees cut into solid lava bedrock and by the remnants of former relief (fig. 3). Many of these scabland channels which formerly carried such large volumes of meltwater carry no streams at present, while others still carry permanent or intermittent streams. Rock-bound lakes are of common occurrence where the coulee floors have been deeply excavated and scoured (fig. 4). Other lakes have been formed in the main valleys dammed by delta deposits, in oxbows on aggraded floodplains of some of the larger streams, in plunge basins at the base of abandoned rapids and falls, and in still other types of depressions on the valley floors (figs. 5, 6). A few ponds lie outside of scabland channels in shallow basins eroded on the basalt plateau.

The age and correlation of the pollen-bearing sediments in the Channeled Scablands of the Columbia Basin of eastern Washington are to be determined largely by the relation of their sites to deglaciation farther to the north. Whereas the last glacier encroached upon the Columbia Basin in only two sectors, the beginning of organic sedimentation in the Channeled Scablands was controlled by deglaciation farther to the north, as these channels carried the meltwater. If, as Bretz (1923) thought, the eastern scabland channels last drained a glacier equivalent to pre-Wisconsin or early Wisconsin, and were not used during the later moraine-building Wisconsin advance, then the sedimentary columns south and southwest of Spokane possibly extend back into late-glacial time. If, on the other hand, the Spokane glacier is of late-Wisconsin age, and its meltwater drained through scabland channels, then the pollen-bearing profiles did not have their beginning prior to the abandonment of these channels. Any pre-Wisconsin or early-Wisconsin sediments that may have existed in these channels were probably removed by late-Wisconsin meltwater, and are not included in the postglacial sedimentary columns.

Hobbs (1943) has attributed the scabland topography to erosion by an ice lobe which extended well beyond the limits of the Spokane lobe.

As the ice retreated from the Columbia Basin and the Columbia and Spokane Rivers carried more and more of the meltwater, the scabland channels separated by the highest divides were presumably abandoned first. Eventually, as the Columbia and Spokane Rivers carried all of the meltwater from the last deglaciation, the scabland channels and streams assumed their postglacial magnitude and other features. Any permanent ponding of water that occurred during this late glacial adjustment provided the sites for potential organic sedimentation from

Bureau of Reclamation photo, Coulee Dam, Washington

FIG. 3.   Channeled Scablands in the Columbia Basin of eastern Washington.   Channel in background carries permanent stream.

Bureau of Reclamation photo, Coulee Dam, Washington

FIG. 4.   Dry Falls Lake, Grand Coulee, Washington.   A rock bound lake.

Bureau of Reclamation photo, Coulee Dam, Washington

FIG. 5.   Dry Falls Lake, Grand Coulee, Washington.   Tule swamp to the right.

Bureau of Reclamation photo, Coulee Dam, Washington

FIG. 6.   Shallow, alkaline lakes in the Upper Grand Coulee, Washington.

early postglacial time to the present. It seems reasonable to suppose that organic sedimentation began here somewhat prior to that in regions farther to the north that were covered longer by the ice. The Okanogan ice lobe which diverted the Columbia River through the Grand Coulee on the western border of the scablands may have persisted for some time after the scabland channels were abandoned by meltwater from the Spokane lobe. While the movements and maxima of the Okanogan and Spokane lobes were not necessarily concurrent, their genesis from the Cordilleran center indicates their probable general contemporaneity with respect to their major oscillations. In view of the possibility that the scabland channels were abandoned at an early stage of deglaciation, they may hold the oldest pollen-bearing sedimentary columns of the Pacific Northwest.

WILLAMETTE VALLEY

Basins favorable for the accumulation of pollen-bearing sediments are comparatively rare in the Willamette Valley of western Oregon, largely because of the absence of Pleistocene glaciation. Although this area was not affected directly by any continental ice sheet nor covered by glacier lobes from the mountains, the valley lowland was inundated by glacial meltwater. During the retreat of the last glacial stage from the hydrographic basin of the upper Columbia River in Washington and Idaho valley fill and numerous ice jams in the Columbia River caused surges of water to back into the Willamette Valley (Allison, 1935). The backwater reached an altitude of at least 400 feet above sea level, or almost 200 feet above much of the present valley floor, as revealed by the position of ice-rafted erratics on the slopes. The composition of some of these erratic boulders indicates that they could have come only from the upper Columbia River basin. During the Pleistocene, also, meltwater from the valley glaciers in the Cascade Mountain Range to the east deposited silts, sands, and gravels chiefly along the eastern border of the valley. The topographic and stratigraphic relations of these deposits suggest alternate stages of alluviation and valley deepening that may tentatively be correlated with the main glacial stages in the· Mississippi Valley (Thayer, 1939; Allison, 1936). These deposits and the erratic-bearing valley fill form a local time scale. The lower Willamette River and its tributaries have excavated their present valley bottoms in the Pleistocene fill and locally have braided or meandering courses. This has resulted in the formation of sloughs and oxbows, as well as other types of depressions on their floodplains. In the course of this erosion downcutting in the main channels caused abandonment of the higher channels. Some of these abandoned channels were blocked at their lower ends by aggrading streams. Such a course of events is illustrated by Lake Labish (map 1). Here water

from the Willamette River once flowed through the Labish channel and thence by the present route of the Little Pudding and the Pudding Rivers (Mt. Angel quadrangle). When the Willamette cut its present channel well below the divide at the head of Lake Labish, a slough lake was left in the abandoned channel, ponded in part by fill from the tributaries. Since then, about 20 feet of pollen-bearing sediments have accumulated in the basin (Hansen, 1942b). Other sites of lacustrine deposits, both organic and inorganic, occur in blocked tributaries of the main streams and on the floodplains of the larger streams where fluctuating water levels during the late-glacial and early postglacial time made complex changes in the drainage. Most of these deposits are shallow and composed of silty muck and partly oxidized organic sediments containing little or no pollen.

The general age of the organic sediments in the Willamette Valley can be dated indirectly as postglacial. The erratics, ice-rafted down the Columbia River from northeastern Washington and Idaho, probably represent the last or late-Wisconsin glacial stage. It is assumed that this stage was chronologically equivalent with the later Vashon stage in the Puget Sound region. Allison (1935) believes that the erratics are equivalent in age to the Vashon glacial deposits of northwestern Washington, as shown by the relative positions of the Willamette Valley fill, erratics, and recent alluvium. If the last glacial stage in the Pacific Northwest is late-Wisconsin in age, all the pollen-bearing profiles located in the Puget Sound region, in northern Washington and Idaho, the Channeled Scablands, and the Willamette Valley are in general equivalent chronologically. It does not follow, however, that sedimentary columns lying upon drift from the last glaciation or its chronological equivalent are exactly of the same age. Nor can it be assumed that organic sediments accumulated on glacial drift in the same region are precisely contemporaneous. Differential rates of lobate advance and retreat, dead ice left by the retreating glacier, ponding of meltwater, use of valleys by streams draining the glaciers, local differences in climate, and many other conditions attendant to the recession of ice undoubtedly provide a varying time range for the beginning of sedimentation even in a relatively small area. If pollen was available and those initially accruing sediments were receptive to pollen and of such characteristic as to preserve it, a time differential of perhaps a thousand years or more may be involved in the record of late-glacial or early postglacial vegetation.

OREGON AND WASHINGTON COAST

Although the coast of Oregon and Washington south of the Olympic Mountains was not glaciated during the Pleistocene, the larger and deeper lakes a few miles inland may owe their origin indirectly to glaciation. The northwest Pacific Coast has had a complex geologic history, involving a number of topographic adjustments during the Pleistocene and Postglacial. Since the development of a series of Pleistocene terraces (Diller, 1896, 1901), these adjustments include Pleistocene and postglacial eustatic changes in sea level, coastal uplift and warping, trenching of marine terraces by streams flowing across them, later drowning and partial filling of these valleys, alternate cutting and filling, deposition of alluvial fans and related deposits, prograding of beach sands, retrograding of shoreline, landward migration of shore dunes, and still other events arising from the interplay of diverse forces and agents. Although there have been apparent movements of subsidence and of elevation of the coast line from time to time during the late Pleistocene and Postglacial, the main change seems to have been an eustatic rise of sea level during deglaciation as a result of the release of water formerly incorporated in the ice sheets. It is estimated that during the last glacial maximum a water layer some 250 feet (Daly, 1929, 1934) to 300 feet (Antevs, 1928) in thickness was removed from the oceans. If the ocean floor, in response to this unloading, rose about one-fifth of the thickness of the removed water layer (Daly, 1934), the actual lowering of the ocean was some 200 to 240 feet. During this low stand of the sea the streams incised themselves to considerable depths so as to maintain accordant junctions. During deglaciation a considerable rise in sea level is denoted by the drowning of river mouths for many miles inland and the extension of submerged channels beyond the mouths of coastal streams (Smith, 1933). The Columbia River valley is drowned for a distance of 140 miles inland, the Umpqua for 25 miles, and the Coquille for 30 miles inland (Fenneman, 1931). The absence of submarine channels extending outward from the mouth of the Rogue River, and the progressive increase of drowning of river mouths both to the north and south, suggest that diastrophism as well as changes in sea level due to glaciation has been at work. The Pleistocene and postglacial physiographic history of the Washington Coast from the mouth of the Columbia River to Grays Harbor, the source of some of the peat profiles of this study, is similar.

The origin of pollen-bearing sediments of the north Pacific Coast probably is related to the late Pleistocene and postglacial sequence of geologic events affecting the area. The development of sites favorable for hydrarch succession and the accumulation of organic sediments have been caused largely by the progress of a marine cycle of a shoreline submerged during the wasting of the continental glaciers. The oldest peat deposits of this cycle are, perhaps, those accumulated in tributaries of main streams that were ponded by valley fill when the sea level began to rise. These lakes occur a few miles inland and have not been affected to any great extent by subsequent sand movement and other processes of the physiographic cycle. Their shorelines are often of dendritic pattern, and it is in these bays and inlets that hydrarch succession and peat deposition has occurred chiefly as a result of shoaling of the water by deposition of sediments from incoming streams, the former tributaries of the drowned main streams. The pollen-bearing sediments in these lakes are the thickest known on the coast, and probably represent continuous deposition from the late Pleistocene or early postglacial time to the present. Undoubtedly many other ponds were formed as early or earlier in the sand dune zone, and hydrarch succession progressed concurrently and substantial amounts of organic sediments were deposited. Periodic shifting of the sand, however, probably buried, drained, or otherwise obliterated these sites. Present day interment of bogs and forests and filling of ponds by dune movement suggest similar events in the past (fig. 8). Indeed, exhumation of bogs buried at some uncertain past date is taking place. An excellent example of this may be seen at the north end of Nye Beach, near Newport, Oregon, where a stratum of peat inclosed in terrace sands has been exposed in a receding sea cliff (Hansen and Allison, 1942). The peat layer is overlain with about 40 feet of sand and may be of late Pleistocene or early postglacial origin. Plant remains in the peat bed show that hydrarch succession had progressed to a climax bog stage which supported arboreal growth.

Many ponds and lakes occupy small depressions between or in front of dunes along the coast. Some of these depressions arise through deflation, while others are formed by complex dune movement. They are continually being formed, obliterated, and reformed by shifting sand. Some of the basins are filled before hydrarch succession gets under way, while others may support an advanced stage of organic accumulation before they are overtaken by advancing dunes. Some of these dunes attain a height of one hundred feet or more, and lakes of considerable depth may be ponded, especially in stream valleys on the leeward side. If sand movement is retarded for a few years, a vegetation cover may stabilize the dunes, and even forests may gain a foothold. The age of organic sediments that persist in ponds of this genesis is uncertain. Individually they probably represent geologically short periods of time. These depressions may have been formed most prolifically at certain relatively dry climatic stages of the past when sand movement was increased by changes in the direction and velocity of winds, the precipitation, and vegetation.

In addition to small ponds formed in these sand dune depressions, extensive peat deposits, usually long and relatively narrow, with the long axes parallel to the seashore, are present in certain areas. These occur chiefly along the northern coast of Oregon and

Photo by U. S. Forest Service

FIG. 7. Stand of Sitka spruce and western hemlock near Otis, Oregon, a few miles from the ocean.
An understory of hemlock has developed as a result of the opening.

Photo by U. S. Forest Service

FIG. 8. Dune sand encroaching upon lodgepole pine forest near Siltcoos Lake, near the Oregon Coast.

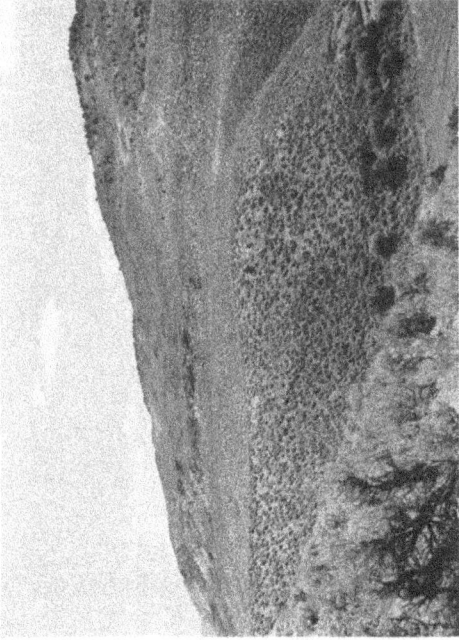

Photo by I. S. Allison

FIG. 10. Slopes on west side of southern part of Warner Basin near site of sedimentary column. Predominant vegetation is sagebrush, with juniper on upper slopes.

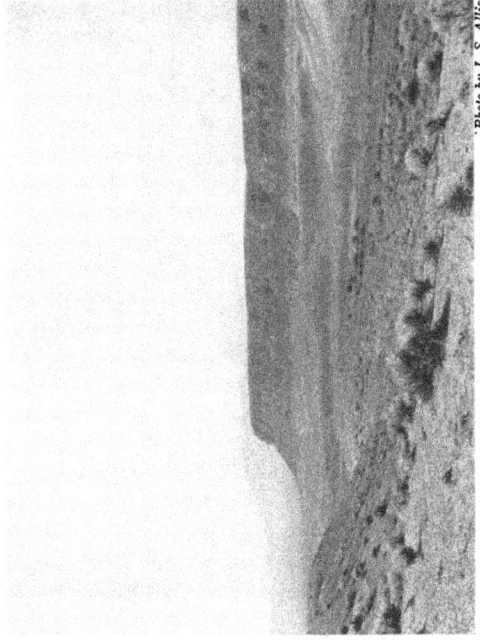

Photo by I. S. Allison

FIG. 12. A view of Warner Basin near Plush, showing dissected early-Wisconsin fill middle right.

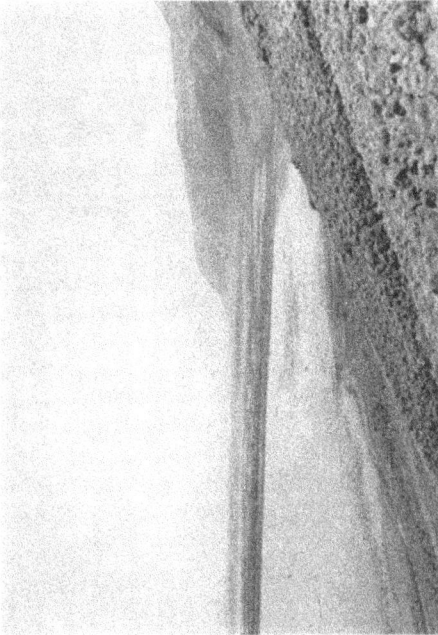

Photo by I. S. Allison

FIG. 9. Warner Basin looking north. Hart Mountain, a fault block, on right showing early-Wisconsin high water level. Playa lake on left.

Photo by I. S. Allison

FIG. 11. Dry bed of Silver Lake, with fault blocks in distance.

on the Washington coast between the mouth of the Columbia River and Grays Harbor. In some areas there is a series of two or three of these elongated peat bogs, separated by high ridges of sand. This type suggests the formation of successive offshore bars and lagoons, or of a succession of parallel beach ridges and associated dunes. If the depressions represent former lagoons, their freshening might require slight uplift above sea level before the hydrarch succession could begin. On the other hand, if the ridges represent successive beaches on a prograding shoreline, no land movement need be involved. In the absence of other evidence the simpler explanation is preferred. In any event, other ridges developed on the shoreward side and inaugurated hydrarch succession and peat accumulation in the accompanying elongated swales. This cycle may have been repeated several times. Subsequent prograding and dune formation have built up a high sand barrier between the present shore and the last-formed swale.

The age of the coast bogs is obviously less certain than that of bogs lying directly upon glacial drift. The oldest living bogs are those that have developed about the margins of the larger and deeper lakes formed in drowned valleys at some distance inland from the shore and beyond the active sand dune zone. It is possible that these organic sediments date back to late Pleistocene when the rise of sea level first began. It does not seem probable that peat deposits in the dune zone are of such antiquity, as the periodic shifting of sand certainly must have buried lakes in all stages of hydrarch succession time after time. Concerning the extensive, elongated peat deposits, one can say only that the original swales are progressively younger toward the present shoreline, but complications from the shifting of sand may be expected. One way of estimating the relative ages of coastal bogs is by the thickness of the organic sediments. As most of the peat profiles are somewhat similar in composition, structure, and degree of compaction, their relative thicknesses should be fair criteria for determining their relative ages. The composition of the adjacent forests during initial sedimentation as recorded by the pollen may shed some light on this problem, as will be shown later.

CASCADES OF WASHINGTON AND OREGON

There are many lake basins in the Cascade Range, most of them comparatively small. By far the greater number owe their origin to Pleistocene mountain glaciation. These include cirque lakes, kettle lakes, recessional and terminal moraine lakes, lakes in ice-scoured depressions, and those formed by obstruction of previous drainage by glacial deposits. Montane glacial lakes are as a rule shallower than those occurring on drift from the continental ice sheet, and their organic sediments are likewise thinner. At least two montane bogs comparable in depth to those of lower elevation have been encountered. Not as many montane lakes and subalpine bogs have been investigated as desired, chiefly because of their inaccessibility and their widely scattered occurrence. In general the organic sediments seem to be thin or lacking in many of the higher lakes, possibly for several reasons that will be discussed later. In addition to glacial lakes, other montane lakes include oxbows, other floodplain types, those dammed by lava flows, landslide lakes, beaver-dammed lakes, and caldera lakes. Peat profiles of this study have been obtained from all of the above mentioned types except the caldera lake. Both Crater Lake and Paulina Lake in the Cascades of Oregon have been examined, but hydrarch succession has not advanced beyond the submerged stage. In addition to bogs and swamps formed in standing water, meadows are numerous at higher altitudes. Most of these occupy broad, shallow basins and contain thin strata of sedge peat with a high fraction of inorganic matter. They usually become hard and dry during the dry summer after the moisture from the winter snow has drained away and evaporated.

Whereas most of the montane lakes in the Cascade Range are of glacial origin, only five of the sixteen profiles classed as montane have been obtained from bogs developed in glacial lakes. Two or three profiles grouped with those from other physiographic provinces are also montane as far as altitude is concerned, but seem to be more nearly related to other groups with respect to development and recorded vegetation history.

At least three stages of Pleistocene mountain glaciation have been recognized in parts of the Cascades. Mackin (1941) believes that the last mountain glaciation in the Snoqualmie-Cedar area of the Cascades adjacent to the Puget Lowland of western Washington was contemporaneous with the Vashon (Wisconsin) stage. The greatest Vashon extension of the local glaciers occurred during an early phase of the Puget glacier maximum called the "Enbankment" stage, while a minor readvance or retreatal stand of the local glaciers, suggested by hilly moraines in the upper valleys of the Cascades, may correspond to the later "Sallal" substage of the Puget glacier. On the east slope of the Cascades in central Washington, in the vicinity of one of the profiles of this study, Page (1939) found evidence for three glacial stages, each of which was less extensive than the preceding. Page says that the third glaciation is undoubtedly equivalent to all or part of the Wisconsin stage. Mackin correlates the late Vashon stage on the west slope of the Cascades with the Stuart or final stage, and the early Vashon stage with the Leavenworth or middle stage as defined by Page. On the east flank of the Sierra Nevada, at least four stages occurred, the McGee, Sherwin, Tahoe, and Tioga, of which the Tahoe

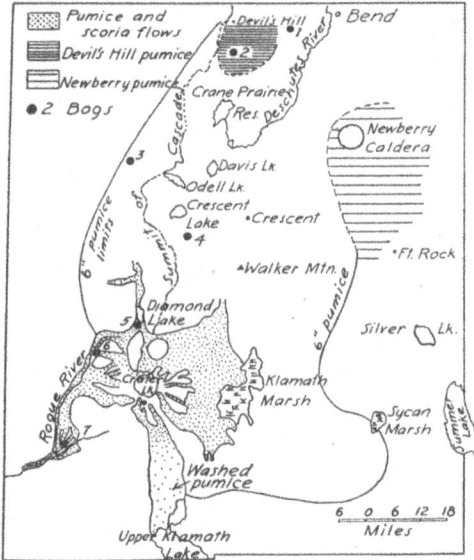

MAP 2. Southern Oregon Cascade region, showing distribution of Mount Mazama pumice and location of bogs overlying or containing an interbedded stratum of the pumice. (From Williams.)

is thought to be equivalent to the maximum of the Wisconsin stage (Iowan), and the Tioga to the late-Wisconsin (Mankato) (Blackwelder, 1931; Allison, 1935). In the Cascade Range of Oregon, Thayer (1939) found three glacial stages which he named the Mill City, Detroit, and Tunnel Creek stages, and tentatively correlated with the Sherwin, Tahoe, and Tioga stages of California respectively. It can be assumed that the fluctuations of the Pleistocene mountain glaciers in this region were more or less synchronous with those of the continental ice sheet. Those sedimentary columns that accumulated in lake basins of glacial origin probably contain the pollen record of the adjacent vegetation for most or all of postglacial time. Those located at higher elevations may have had a slightly later origin, owing to the persistence of mountain glaciers for some time after the dissipation of the continental ice sheet. In Washington the occurrence of the postglacial volcanic ash stratum in several montane bogs, common to most pollen-bearing profiles in Washington, indicates that they are nearly as old as those at lower elevations within the glaciated region. A bog located near Wenatchee, Washington, at an elevation of 4,200 feet was apparently formed in a landslide basin of unknown age. The occurrence of the volcanic ash stratum, however, aids in its correlation with other bogs containing the ash.

Several of the montane bogs of this study lie upon the great pumice mantle distributed by the eruption of Mount Mazama in southern Oregon (Williams, 1942). Other sources of pumice are Newberry Crater and several other cones in the Three Sisters area (Williams, 1935, 1944). The distribution of Mount Mazama pumice has been mapped by Williams, and most of the pollen-bearing sediments in this region rest upon it (map 2.) At least one sedimentary section rests upon pumice from another source, probably Devil's Hill, located about 24 miles west of Bend. Pumice from the Newberry Crater, which lies slightly east of the main Cascades, was not dispersed very far to the west. The eruption of Mount Mazama occurred after the late-Wisconsin maximum, and bogs resting upon its pumice are necessarily younger than those resting directly upon glacial drift.

NORTHERN GREAT BASIN

Finally, a type of lake basin that cannot be placed in any of the foregoing categories is that of the Great Basin. This physiographic province extends into south central Oregon and is characterized by a series of fault block mountains and broad, shallow depressions with largely interior drainage (fig. 9, 11). During the Miocene the Columbia lava flows covered most of this region, and continued elevation and considerable faulting took place until late Pliocene or possibly early Pleistocene. During the Pleistocene the moister climate caused lakes to form in the depressions which since have undergone considerable fluctuations in response to climatic cycles, as is shown by the presence of old shorelines. The smaller and shallower lakes have become filled, desiccated, or reduced to playas (fig. 11). Some of the larger lakes have become extremely alkaline so that little or no vegetation thrives, while others support various swamp stages about their margins or have become entirely filled with organic sediments. The largest lakes of the northern Great Basin include Harney, Malheur, Alvord, Warner, Goose, Abert, Summer, Silver, and Upper and Lower Klamath (map 1). Silver, Harney, Goose, and Lower Klamath lakes are either dried up or drained. The Warner Lakes consist of a chain of several small lakes that are remnants of a few large ones. Other basins also have become filled with organic sediments and are now marshes. These include Chewaucan Marsh, formed in a remnant of the Pleistocene antecedent of Summer Lake, Sycan Marsh, and Klamath Marsh.

During the Pleistocene and early Postglacial, many of the Great Basin lakes were much higher than today, as is shown by one or more well-preserved shorelines (fig. 9). These high water stages of these Pleistocene lakes are probably to be correlated with the maximum stages of the Pleistocene glaciers in the western mountains and with the continental ice sheets in the north, although they may be slightly later. Antevs (1938) correlates the highest stages of lakes Bonneville and Lahontan with the Iowan-Tahoe glacial stage, and the Provo shoreline in Bonneville basin and the Dendritic terrace in Lahontan basin with the Mankato-Tioga glacial stage. In Oregon, Allison (1945) has correlated the highest level of Lake Chewaucan, the Pleistocene antecedent of Summer Lake, with the Tahoe (Iowan or early Wisconsin) glacial stage of the Sierra Nevada, and the lower beach levels of Winter Lake with the Tioga (Mankato or late Wisconsin) stage.

Although the total area represented by pollen dispersal to the sites of organic sediments of this study is only a small part of the entire Pacific Northwest, the only phytogeographic area not represented is southwestern Oregon. The mature topography of this region and absence of standing water, however, has prohibited the accumulation of pollen-bearing sediments. None was located on trips into this area, while examination of topographic maps reveals the virtual absence of lakes, bogs, or swampy areas that might be the sites of organic sedimentation.

SUMMARY

Most of the lake basins which have supported postglacial hydrarch succession resulting in the formation of pollen-bearing sediments in the Pacific Northwest are related, directly or indirectly, genetically to Pleistocene glaciation. In the Puget Lowland and northern Washington and Idaho, the depressions were left in the wake of the retreating ice. In the Cascade Range of Washington and Oregon lakes were ponded in basins owing their origins to mountain glaciation. Beyond the limits of glaciation depressions supporting hydrarch succession were formed as a result of various geomorphic cycles that were initiated and controlled by the influence of glaciation. These include inundation by glacial backwater by streams heading at the ice front, shoreline processes initiated by eustatic changes in sea level in response to nurturing and wasting of the ice sheet, the cutting of channels by glacial meltwater and subsequent abandonment leaving ponded water on certain sites as in the Channeled Scablands of eastern Washington, aggrading of main stream valleys resulting in blocking of tributary mouths, and yet other erosional and depositional processes.

Some pollen-bearing sediments have accumulated in depressions that have no relationship to glaciation. These include landslide depressions, caldera lakes, fault basins, lava-dammed streams, beaver dams, and oxbow lakes. The time of their formation is difficult to estimate.

A few basins and their sediments may date back to glacial time, and some of them are undoubtedly much younger than postglacial. Most of them, however, are considered to be postglacial in time of origin. The lapse of time between deglaciation in each region and the beginning of sedimentation in each basin is unknown. The time differential between deglaciation in the several regions covered with Pleistocene glaciers is also unknown, and such estimates can be only approximate and relative.

POLLEN–BEARING PROFILES OF THE PACIFIC NORTHWEST

GENERAL STATEMENT

The deposition of organic sediments is an exceedingly complex process, involving a compromise of the reactions and coactions of the physical, chemical, and biological environment. These factors, both intra- and extra-aquatic, are so interrelated that often the limiting one is obscured and most of the others defy measurement. The depth, composition, structure, and typological succession of peat profiles are an expression of the total impact and interrelationship of the several phases of the environment during their deposition. The trends in environmental changes are reflected by the physical and chemical structure and the typological succession of the sediments, and some of these trends can be interpreted readily from the sedimentary column. Peat profiles that have accumulated under different environmental conditions likewise record these variations. The aquatic environment is probably more constant than the terrestrial, and typological succession of the sediments is more a result of modification of the aquatic biocoenose itself rather than that of the adjacent non-aquatic habitat. The pollen profiles, however, are excellent examples of alterations of the extra-aquatic environment that are recorded in the sedimentary column. Dachnowski (1926) and others have interpreted climatic trends from the succession of peat types. This may be feasible in some areas, but it would seem that the limiting factors in determining the type of organic sediment would have to be at a critical minimum in order that changes in the extra-aquatic environment be indicated by significant changes in sediment types. In order to interpret climatic trends from peat profiles, the indicator strata would have to be consistently present, and in the same relative stratigraphic position in several profiles within a homogeneous physiographic, climatic, and floristic province. Such interpretations have been widely applied to European peat profiles with considerable success (Sears, 1942).

The wide geographic range of the sedimentary basins of this study, with a correspondingly wide variation in climate, physiography, and flora, has resulted in several types of peat and other pollen-bearing sediments. In speaking of a sedimentary column composed of both inorganic and organic pollen-bearing strata the term pollen-bearing sediments is applied. The term peat implies an organic, fibrous material, with a low fraction of inorganic content, and the term peat profile does not hold for some of the sediments of this study. Pollen grains may be present in the sand and even fine gravel underlying the finer sediments, as well as in silts and clays occurring progressively upward in most profiles. The term bog for all of the areas of hydrarch succession involved in this study seems undesirable because some of the pollen-bearing sediments have accumulated in swamps or marshes rather than in bogs. The sediments in swamps are often composed of muck and contain a high fraction of inorganic matter. There are various definitions of bogs and swamps, and there is probably no sharp line of demarcation between them. It seems reasonable to define a bog as a stage of hydrarch succession where sediments are composed almost entirely of organic materials, and where the water table rarely rises above the surface. The latter condition limits the income of inorganic sediments by erosion, although it does not prevent the entrance of wind-borne inorganic material such as volcanic ash, pumice, silt, and sand in sand dune regions. Swamps are usually formed in more extensive and shallower basins where seasonal flooding of the surface carries in considerable silt and clay. These inorganic materials are commingled with the plant remains and form a dark, mucky sediment. The seasonal flooding and fluctuation of the water table throughout the year result in more oxidation and decay of the organic remains. In a true bog the peat usually has sufficient absorbing capacity to hold the incoming water during the wet season without being inundated. During the dry summers, such as occur in the Pacific Northwest, the capillarity of the peat carries water to the upper part of the substratum for use by the living plants. Swamps are often better drained than bogs, and lack of capillarity in the substratum causes more of a seasonal aspect in the flora due to variation in the amount of available water. A bog may develop on a swamp as hydrarch succession progresses and modifies the environmental conditions.

UNDERLYING MATERIALS

There is a variety of materials underlying the pollen-bearing sediments of this study. They include surficial glacial till, glaciolacustrine, glaciofluvial, eolian, fluvial, volcanic, and possibly marine deposits, and bedrock. In boring the sedimentary column an attempt has been made to obtain samples from the lowest and earliest pollen-bearing sediments in order to have as complete a record as possible of the post-Pleistocene forest succession. In most sections pollen is scarce in these underlying strata. Potzger and Wilson (1941) have shown the importance of obtaining samples of the sedimentary column to some depth beneath the organic sediments in northern Indiana.

By a special device they have obtained samples of several feet of underlying sand, and found pollen grains abundant, thereby extending the forest record of the earliest period. The underlying strata of sand in the pollen-bearing sediments of the Pacific Northwest have been studied and very little pollen has been noted. Where sufficient pollen grains were present, their proportions have been included in the pollen profiles. In a bog near Sedro-Woolley, Washington, a stratum of silt and clay, 2.25 meters thick, was examined and no pollen grains were noted, suggesting absence of forests in the region at the time of deposition. Instead, a varve-like structure of this inorganic stratum denotes the proximity of the glacier. Examination of half-meter of underlying sand from sand dune bogs on the Oregon coast reveals the almost complete absence of pollen grains as well as other organic matter.

Pollen grains are also absent from pumice and volcanic ash underlying the organic sediments of bogs in the Cascades and Blue Mountains of Oregon. The pollen-bearing sediments in the scablands of eastern Washington rest directly upon the bedrock of the Columbia Basin, suggesting presence of vegetation in the region at the time of earliest sustained deposition. The presence of sand particles throughout the peat profiles in the sand dune zone of the Oregon coast indicates continuous sand movement.

TYPES OF SEDIMENTS

The inevitable fate of lakes and ponds is obliteration by filling of their basins. The rate of sedimentation and the time required to fill lake basins are determined by their size and depth, physiographic setting, lithology of adjacent terrain, climate, chemical composition of their water, type of vegetation, and other factors. One important process is filling by organic sediments derived from hydrarch succession. This process involves the rise, culmination, and decline of successive groups of organisms as a result of the modification of the aquatic environment, usually by the organisms themselves. It begins with the earliest submerged stages composed chiefly of plankton and continues up through the final stage by which a substratum supporting the climax biota of the region is produced. Often all stages are present in a single lake, arranged in roughly concentric zones from the center outward. Each stage contributes the remains of its dead organisms to the accumulating sediments along with the organic and inorganic material which may be carried into the basin by wind and water. The typological succession of the sediments records the transition from stage to stage, from the silty organic deposit at the bottom to the brown fibrous peat of the bog or the mucky peat of the swamp or marsh at the top.

The rate of deposition, kind of sediments, and typological succession in a lake are due in part to the lake type as based upon its productivity. Lakes have been classed as three major types, depending upon their biological productivity: viz., oligotrophic, eutrophic, and dystrophic (Welch, 1935). These types are determined largely by the dissolved nutritive material present, the oligotrophic being poor and the eutrophic being rich in these substances. In the dystrophic type the presence of the humic materials seems to be a more important factor in limiting development than the nutritive materials. These lake types are not permanent and may be converted from one type to another as hydrarch succession progresses or retrogresses. Usually, however, a definite sequence is involved. Strøm (1928) says that the natural process in the maturing of a lake is eutrophication. Apparently many lakes are oligotrophic to begin with, and gradually become eutrophic as the amount of organic nutrients is increased. When the littoral vegetation gains a foothold owing to shoaling as a result of sedimentation, and brown peat is formed, the lake is converted into a dystrophic lake. Thienemann (1931) states that the oligotrophic is succeeded by the eutrophic lake which in turn develops into a pond, swamp, or moor. The dystrophic lake is converted in to a peat bog. Most of the lakes and ponds in which the sedimentary columns of this study have accumulated, apparently, were initially of the oligotrophic type.

In most Pacific Northwest postglacial sedimentary columns, formed in lakes, there is no sharp line of demarcation between the inorganic and the organic sediments. There is usually a gradation from the coarser inorganic sediments up through the finer, with an ever-increasing proportion of organic detritus until there is little or no mineral matter present. The gradation from purely inorganic to almost purely organic involves varying thicknesses of sediments in different bogs. The characteristic sediment overlying the inorganic material is usually a gelatinous or greasy, colloidal, finely divided substance varying in color from tan or yellow, through gray, and into many shades of green. In some cases it is of cheese-like consistency. It is composed largely of autochthonous materials derived from pelagic plankton organisms, the bodies and excreta of benthic animals, and to some extent the fragments of littoral higher vegetation. The amount of allochthonous materials is limited although erosion may carry the remains of terrestrial organisms into the bond before a zone of higher littoral vegetation develops about the margin. The pollen grains from the adjacent forest trees represent an important component of the allochthonous sediments. This type of sediment is apparently equivalent to "gyttja," a term first introduced by von Post (1862) and applied to similar sediments in Sweden. He defines it as an organic sediment formed in lakes with clear and nutritious water, or in the bays of the sea, from organic detritus (algae and flowering plants; aquatic insects, mollusks, excrements, etc.), clay, and silt. In Sweden the colors are usually brown-green, gray-green, or green of various shades. Lindeman (1941) and Deevey (1942) have adopted this term for similar sediments in the Great Lakes region and New England. Other terms applied to this type of sediment are "limnetic organic ooze" (Auer, 1930), sedimentary peat (Rigg, 1940a), and "limnic" peat (Voss, 1934). Some authors have used the terms muck, slime, and mud without further explanation. Granlund (1936) describes several types of gyttja, including shore-gyttja, plankton-gyttja, clay-gyttja, calcareous gyttja, and others. Dacknowski-Stokes (1936) refers to what is apparently the equivalent of gyttja by the terms sedimentary muck, diatomaceous muck, and olive-green organic sediments.

The characteristic deposit of the dystrophic lake is "dy" which is translated into peat or lake peat. This term was also introduced in scientific literature by von Post (1862) who defined it as an organic sediment formed in lakes with brown water. It is rich in humus from colloidal humus and organic detritus. The chemical decomposition takes place during partial exclusion of air. The color of dy is dark brown, quickly changing to black when exposed to air. Apparently the chief difference between gyttja and dy is that in the former the organic matter is largely decomposed by putrefaction, while in dy it is largely colloidal humus. Granlund (1936) describes lake-dy as consisting chiefly of colloidal humus transported from the land surrounding the lake. The color is dark-brown, but quickly turns black in the air. The consistency is even and grainy. This type of organic sediment is not common in Pacific Northwest profiles. A sediment called brown sedimentary peat by Rigg and Richardson (1938) is probably equivalent to dy. The author has found a dy-like sediment in sand dune bogs on the Oregon coast. In these bogs the characteristic gyttja stratum is absent, and the underlying sand grades directly into dy and then fibrous peat. A somewhat similar typological succession has been noted for sedimentary columns resting upon pumice in the Oregon Cascades.

HYDRARCH SUCCESSION

Most of the higher plants concerned with hydrarch succession are cosmopolitan and are widespread throughout Northern United States and Canada. These include plants of the submerged stage up through the bog and swamp stages. The more mesophytic vegetation that invades the surface of bogs and swamps in their climax stages, however, consists of the species that are a part of the climax of that particular region. The true bog plants, associated with *Sphagnum* and high acidity of substratum, are mostly boreal relicts that have persisted in favorable habitats since they migrated south before the advancing Pleistocene glaciers (Transeau, 1903). Most of the submerged, floating, and swamp plants are less exacting in their ecological requirements, and the glacial influence upon their distribution was largely topographic. Many species of the first two series are common to all areas of this study. The bog plants are most abundantly distributed west of the Cascades, and specifically in the Puget Sound region and along the Oregon and Washington coasts. A few species occur in the Cascades at higher elevation and in the northern Rocky Mountain section of Washington and Idaho. The greatest digression from the typical bog is in the scablands of eastern Washington. The climate and alkalinity of the substratum have probably been the chief limiting factors in determining the hydrarch succession. The dry summers with consequent lowering of the water table, the seasonal flooding, and an alkaline substratum derived from the basalt have not provided the optimum conditions for bog development. Some of the lakes are too alkaline for hydrarch succession, particularly those located farthest down-valley in the coulees. The alkalinity increases down-valley owing to dissolution of salts by the water as it percolates down-stream and to progressive evaporation en route. This is shown by the gradual decrease in the number of freshwater plants and the increase in halophytes from lake to lake in some of the coulees. Increase in the alkalinity of the upper substratum in the more mature swamps is shown by the invasion of halophytes. The concentration of salts in the upper sediments has undoubtedly increased from seasonal flooding and evaporation. Alkalinity has probably increased as the swamps reached maturity and the rate of organic sedimentation was retarded in more recent time. The abundance of *Chara* sp. in the submerged stage of two lakes in northern Idaho and Washington, and a thick stratum of Chara marl in the sedimentary column suggest an abundance of calcium available for assimilation by this plant (Hansen, 1940b, 1943e).

PACIFIC NORTHWEST SEDIMENTARY COLUMNS

The occurrence of lake basins in the Pacific Northwest under a wide range of climatic and physiographic conditions has resulted in the deposition of several kinds of pollen-bearing sediments from the standpoint of composition. The great number of higher plants involved in hydrarch succession indicates the complexity of fibrous peat, while an analysis of the limnic peat underlying sediments of higher inorganic fraction would undoubtedly reveal that as many or more organisms have contributed to its makeup. In spite of the diversity of the organisms concerned in organic sedimentation in the several regions, the basic structural types are similar and are here classified as limnic and fibrous peat. Rigg and Richardson (1938) in an intensive study of Sphagnum bog profiles in the Pacific Northwest west of the Cascades distinguish twelve distinct kinds of strata based upon color, structure, and composition of the sediments. They are lake mud, sedimentary peat, sedge peat, tule-reed peat, grayish-brown fibrous peat, reed peat, wood peat, charred remains, heath peat, muck, Sphagnum moss peat, and volcanic ash. The bulk of most profiles is composed of lake mud, sedimentary peat, sedge peat, and Sphagnum moss peat. The lake mud is underlain in most places with clay, and only in a few instances was boring extended down into sand and gravel. Of the twelve kinds of strata listed nine are unmodified organic sediments and can be classified generally as either limnic or fibrous peat. Rigg states that there are some overlapping properties in lake mud, muck, and sedimentary peat. Microscopic examination of lake mud shows that it is basically an organic sediment equivalent to gyttja. The term muck has been used indiscriminately, but the writer uses this term to refer to organic sediments that have been changed by oxidation or decomposition by aeration and micro-organisms. This definition agrees with that of Dachnowski-Stokes (1941). He says that muck is residual peat material, generally dark brown to black in color, and relatively low in absorbing capacity for water and soluble salts. The mineral content may range between 5 and 35 per cent. Continued oxidation obviously increased the proportion of inorganic materials.

Dachnowski-Stokes (1936) in a study of Pacific Northwest peat bogs in relation to land and water resources uses a more voluminous terminology than that of Rigg. In addition to certain of the latter's terms he applies the terms of tule peat, diatomaceous sedimentary peat, sedimentary muck, sedimentary fibrous tule peat, tule muck, sedimentary fibrous sedge peat, sedimentary fibrous muck, sedge muck, sedimentary fibrous cattail peat, sedimentary fibrous peat, Hypnum peat, diatomaceous muck, olive-green organic sediments, woody muck, Equisetum peat, woody peat, woody sedge peat, and still others. Most or all of the above terms referring to unaltered organic sediments can be classified as gyttja, dy, or fibrous peat. In a later paper Dachnowski-Stokes (1941) classifies peat into three types: woody, fibrous, and sedimentary. It is realized that it becomes important to classify organic sediments in greater detail when they are of economic value or have some bearing upon water and soil conservation practices. In studies of prehistoric hydrarch succession and typological succession on the basis of peat type succession, it also is essential to distinguish the many compositional types even though there are only a few structural types.

In previous pollen studies the author has used only

a few terms in classifying the organic pollen-bearing sediments although a detailed record was made for each profile. These terms included sedimentary peat, limnic peat, marl, sedge peat, Sphagnum peat, Hypnum peat, and fibrous peat. Some of these terms were qualified with an adjective to denote color, degree of coarseness, etc. The first three types are equivalent to gyttja or dy while the others are fibrous peat. It is not within the province of this paper to discuss in greater detail the structure and composition of the sediments encountered. They have provided chiefly a means to an end; namely, a medium for preservation of the pollen grains that were interred in the accumulating sediments as hydrarch succession progressed. The principal objective in differentiating the few types of sediments is to provide a more reliable basis for chronology as determined from the estimated rate of deposition. No systematic evidence has been noted for climatic trends in the typological succession of the sediments although the author does not feel competent to translate typological succession into extra-aquatic environmental trends except by interpretation of the pollen profiles. In order to work out the hypothetical chronology of postglacial time in a simple and general way, based chiefly upon the quantitative factor, the postglacial pollen-bearing sediments in the Pacific Northwest are classified as sand, silt, clay, limnic peat, and fibrous peat. They usually occur in this order from the bottom upward. Certain unusual strata, such as marl, diatomite, and volcanic strata, are mentioned or discussed in relation to their value for chronological correlations. Tables 1–7 show the thickness and proportions of the several types of sediments for each profile of this study.

PUGET SOUND REGION

The Puget Sound region supports the greatest number and the maximum development of Sphagnum bogs in the Pacific Northwest. The latter is manifested by the greatest average depth for the pollen-bearing profiles of this region of all the groups involved in this study. In fact, the conditions favoring hydrarch succession in general have been more nearly at an optimum here than in the other regions. These conditions include an abundance of undrained sedimentary basins, annual rainfall sufficient to maintain a fairly constant high water table in the bogs, cool summers, and a long growing season. Most of the bogs have developed in kettles and other types of undrained glacial depressions which have limited fluctuation of the water table and its attendant retarding of plant growth and rate of organic sedimentation. *Sphagnum*, one of the chief contributors to more recent peat strata, has no seasonal dormancy and thus may continue its growth practically throughout the year. Only five of the sixteen bogs of this study in the Puget Lowland are not of the Sphagnum

TABLE 1

SEDIMENTARY COLUMNS FROM THE PUGET LOWLAND OF WASHINGTON

Location	Depth in m.[a]	Thickness overlying volcanic ash	Sand	Silt	Clay	Limnic peat	Fibrous peat
Ronald	6.5	2.3			0.2	1.2	5.1
Parkland	10.5	2.8			0.2	3.5	6.8
Black Diamond	6.0	2.0		1.0		1.0	4.0
Sedro-Woolley	9.5	4.5		2.25		3.25	4.0
Mount Constitution	9.0	7.75			0.5	5.25	3.25
Killebrew	9.5	7.0	0.5	0.5		5.5	3.25
New Westminster	4.25	2.5			0.25	1.5	2.5
Lulu Island	5.0			0.5	0.75	1.5	2.25
Poulsbo	8.75	5.5		0.5	0.25	3.0	5.0
Bellingham	13.0	8.0	0.25	0.75		7.0	5.0
Granite Falls	8.0	5.3	0.25	0.25		4.75	2.5
Olympia	5.4	3.2		0.2		1.2	4.0
Rainier	3.0	1.0	0.2			2.6	0.2
Tenino	13.0	9.6	0.5			6.5	6.0
Silver Lake	3.0				0.2	1.8	1.0
Farger Lake	6.0					4.4	1.6
Average	7.8	4.7				3.4	3.6

[a] All figures in meters.

type. The many floristic aspects of the bogs of the Puget Sound region have been adequately discussed by Rigg (1925, 1934, 1938, 1940a, 1940b), and as they have only indirect bearing upon the problem at hand, further consideration will be omitted.

The average thickness of sixteen pollen-bearing sedimentary columns in the Puget Sound region is about 7.8 meters. They range in depth from 3.0 to 13 meters. An average of about 3.4 meters or 45 per cent of the columns is composed of limnic peat, while an average of 3.6 meters or 48 per cent constitutes fibrous peat. The balance is represented by varying proportions of sand, silt, and clay (table 1). The limnic peat is rather consistent in its appearance and structure in the several profiles, and varies from gray to brown in color. The inorganic fraction decreases, the color darkens, and the amount of fibrous material increases upward in the profile. The chief digression from an otherwise smooth, homogeneous, and cheese-like consistency is the presence in some profiles of a gelatinous material, ranging from yellow to olive green in color. Several feet of this type of sediment occurs in the Killebrew Lake bog, on Orcas Island (Hansen, 1943b). Rigg (1940b) ascribes the source of this jelly-like material to algae. Another modification of gyttja in some profiles is brought about by the occurrence of diatoms which increase in abundance immediately above the volcanic ash stratum. The ash probably provided considerable silica essential to the diatoms in the formation of their tests. In some cases the diatoms are sufficiently concentrated to designate such strata as diatomaceous gyttja.

Sedge and Sphagnum peat are the principal kinds of fibrous peat found in Puget Sound bogs. In their order of deposition the former comes first. In five of the bogs the sedge stage had never been converted into the Sphagnum stage. In a few bogs the sedge stage never developed, and Sphagnum is underlain directly with limnic peat (Rigg and Richardson, 1938). Various theories have been advanced for the absence of this stage, and it seems probable that the principal reason is extreme fluctuation of the water table at certain periods in the development of the bog. It seems doubtful that the absence of a sedge stratum represents an unconformity. The sedge peat may consist of a wide variety of plant remains but several species of *Carex* are the principal contributors. In a few instances it is modified by the presence of reed (*Phragmites*), cattail (*Typha latifolia*) and dicotyledonous plants. Other sedges or sedge-like plants that compose the sedge peat are one or more species of the following genera: spike rush (*Eleocharis*), beak rush (*Rynchospora*), three-way sedge (*Dulichium*), rush (*Juncus*), bulrush (*Scirpus*), sedge (*Cyperus*), burreed (*Sparganium*), and cotton-grass (*Eriophorum*).

Moss peat in the Puget Sound region is composed chiefly of species of Sphagnum although Hypnum moss may locally be the preponderant component. The hairy cap moss (*Polytrichum*) may be prevalent where fire has provided the essential conditions for its growth, and its remains have been noted immediately above charred horizons. Spagnum leaves are often present throughout the limnic stratum, and increase in abundance upward until practically pure Sphagnum peat exists. Where there is a stratum of sedge peat, fragments of *Sphagnum* may be present throughout, the occurrence of which suggests the existence of a Sphagnum mat on other parts of the lake during that period.

The average thickness of the sediments overlying the volcanic ash stratum in bogs of the Puget Lowland is about 4.7 meters. It was not noted in three of the profiles, while in others it does not occur as a discernible layer but as dispersed fragments of glass. The latter condition exists where the ash fell upon bogs in the Sphagnum stage at the time of the volcanic eruption (Hansen, 1940a). In eight of the profiles the volcanic glass stratum occurs in the limnic peat, indicating that it fell on water and settled to the then existing bottom. The ash stratum becomes progressively thinner and less evident in bogs southward and westward from the Puget Sound region until finally it is apparently absent in bogs on the Washington coast and south of Chehalis in the Puget Trough. This suggests that the source of the ash was a volcano in the Cascades of northern Washington. The presence of a layer of volcanic ash on glacial moraines of late-Wisconsin age in this region

and its assignment to Glacier Peak in northern Washington (Waters, 1939) explains its diminution to the west and south.

Charred horizons occur occasionally in the sedimentary columns, and may represent fire on the bog surface itself or charred material that was carried into the bog either by wind or water from adjacent areas. The occurrence of charred materials in the limnic peat precludes the possibility of fire on the bog because of the stage of hydrarch succession. It is possible that fire occurred on parts of the bog surface that were in a mature stage of succession, and the charred remains were then dispersed into the open water. Fires on the bog surface must have resulted in unconformities in the sedimentary column although such unconformities are not reflected in the forest succession as interpreted from the pollen profiles. Changes in pollen abundance of certain species in adjacent areas immediately above the charred horizon indicate the favorable or unfavorable influence of fire upon their existence.

PACIFIC COAST

Conditions favoring bog development have probably been near an optimum on the north Pacific Coast during the post-Pleistocene. The most adverse process limiting sustained peat deposition apparently has been the periodic shifting of sand on such a large scale as abruptly to terminate hydrarch succession by burying a given site. The rainfall is heavier than in the Puget Sound region, and the temperature more equable. Conditions more favorable for the deposition of fibrous peat than for that of limnic peat are manifested by the general absence of the latter in the sand dune bogs. The shallow depth of the lakes permitting almost immediate encroachment by littoral vegetation, as well as the extreme oligotrophy, has perhaps been an important factor in limiting deposition of limnic peat. The species of higher plants and the manner and order of their succession are similar to those of the Puget Sound region.

The average depth of eleven pollen-bearing sedimentary columns on the Pacific Coast is almost 4 meters. They range in depth from 1 to 11.75 meters. An average of about 0.9 meter or 22 per cent is composed of limnic peat, and an average of about 2.3 meters or about 60 per cent is fibrous peat. The disparity between the total average of limnic and fibrous peat and the average depth of the profiles is due to the high proportions of clay and silt in the Woahink Lake and Hoquiam profiles (table 2). Sand is present throughout the dune bog sedimentary columns and is probably of eolian origin. In all but one of the bogs the fibrous peat is composed chiefly of *Sphagnum* and the remains of its associates. No volcanic glass was noted in any of the profiles, as the coast bogs are probably located too far from the sites

TABLE 2

SEDIMENTARY COLUMNS FROM THE OREGON AND
WASHINGTON COAST

Location	Depth in m.[a]	Sand	Silt	Clay	Limnic peat	Fibrous peat
Bandon......	4.4	0.4			2.0	2.0
Marshfield...	2.1	0.1	0.1			1.9
Hauser......	1.0		0.2		0.2	0.6
Woahink Lake	11.75	0.5	0.5	3.75	1.75	5.25
Newport.....	1.0	0.1				0.9
Sandlake.....	4.0	0.2			0.8	3.0
Gearhart.....	7.0		0.25		2.75	4.0
Ilwaco.......	2.8	0.1			0.1	2.6
Grayland....	2.1	0.1			0.4	1.5
Hoquiam.....	4.4		0.4		0.6	1.4
Forks.......	2.5	0.1	0.3		1.0	1.1
Average....	3.8				0.9	2.3

[a] All figures in meters.

of volcanic activity responsible for the glass in other bogs in the Pacific Northwest. Charred horizons are present in several of the profiles, and in some cases the influence of the recorded fire in adjacent vegetation is reflected by the pollen profiles.

WILLAMETTE VALLEY

The scarcity of post-Pleistocene organic sediments in the Willamette Valley, suggests that conditions for hydrarch succession have not been so favorable as in other parts of the Pacific Northwest. The dearth of undrained depressions in conjunction with inadequate precipitation during the summer has probably been the chief factor in limiting the deposition of pollen-bearing sediments. Most of the organic deposits have accumulated under swamp rather than bog conditions, as is indicated by their higher inorganic fraction, the occurrence of muck horizons, and the species of plants involved in hydrarch succession. The mineral matter, composed largely of silt, denotes periodic inundation, while the muck horizons suggest lowering of the water table below the surface, permitting oxidation and decomposition for unknown periods of time.

The rarity of Sphagnum bogs in the Willamette Valley reflects the unfavorable topographic, limnologic, and climatic conditions necessary for their development. The floodplain depressions, sloughs, and oxbows, which have served as the sites for hydrarch succession, have outward drainage. Heavy precipitation during the winter and little or none during the summer result in such great fluctuations of the water table as to inhibit the growth of Sphagnum. Sedges, sedge-like plants, and other plants concerned with hydrarch succession in maturing swamps can withstand both substratal and surface desiccation better than Sphagnum. The rhizomes, tubers, and

roots serve as water-storage organs and also as a means of survival during the unfavorable season. Sphagnum moss dies upward from the bottom, thus destroying the chief means of conveying water from the substratum to the growing portions. If the water table sinks far below the surface, there is little chance for water to reach the living part of the plant. Oxidation and decomposition of the underlying dead Sphagnum tends further to accentuate this process.

The average depth of six pollen-bearing sedimentary columns in the Willamette Valley is 7.0 meters. They range in depth from 3.75 to 12 meters. An average of about 2.5 meters or 37 per cent is composed of limnic peat, while an average of about 3.5 meters or 50 per cent consists of fibrous peat. One of these, located at Wapato Lake, is almost devoid of pollen at most levels and is not included in the pollen profiles. A high average proportion of silt and clay underlying the organic sediments denotes the recurrent flooding in the early stages of sedimentation. About 3 meters of mineral material underlie the limnic peat in the Onion Flats profile (table 3). In two profiles from Lake Labish, the lowest meter is composed of silt almost devoid of pollen. It is included in the average depth, although it is omitted from the pollen profiles because of the paucity of pollen. Both Onion Flats and Lake Labish have been drained and placed under cultivation for about three decades, and considerable subsidence has occurred. At Lake Labish, about 2 feet of subsidence had occurred after two decades of cultivation, and Powers (1932) estimated subsidence at a rate of about 1 inch per year. At least 3 feet of subsidence have probably taken place to date owing to oxidation and decomposition, and further lowering of the surface level has undoubtedly resulted from deflation. Differences in degree of subsidence and loss of surface materials by deflation between Onion Flats and Lake Labish may be shown by the stratigraphic position of a pumice stratum. At Lake Labish a well defined pumice layer occurs at a depth of 1.75 meters, while

TABLE 3

SEDIMENTARY COLUMNS FROM THE WILLAMETTE VALLEY
OF OREGON

Location	Depth in m.[a]	Thickness overlying pumice	Sand	Silt	Clay	Limnic peat	Fibrous peat
Wapato Lake...	4.25		0.25		0.75	2.25	1.0
Noti..........	3.75	2.0	0.1	0.3	0.1	1.75	1.5
Labish No. 1....	7.0	1.85		1.0		2.0	4.0
Labish No. 2....	7.0	1.85		1.0		2.0	4.0
Onion Flats.....	12.0	0.5			2.0	4.5	5.0
Scotts Mills.....	8.4			0.2	0.2	2.5	5.5
Average......	7.0					2.5	3.5

[a] All figures in meters.

at Onion Flats volcanic glass, presumably the equivalent of the pumice, occurs between 0.5 and 0.75 meter. This disparity in the position of the glass may be the result of longer cultivation of Onion Flats and/or other factors. The source of the pumice is unknown as there has been considerable post-Pleistocene volcanic activity in the Oregon Cascades. It did not come from the eruption of Mount Mazama in southern Oregon.[1] It may have been derived from Mount Saint Helens in southern Washington.

Two of the sedimentary columns were taken from Sphagnum bogs. One is located near Noti in the foothills of the Coast Range, where the precipitation is heavier than in the valley proper. Not so many bog plants as occur in the Coast and Puget Lowland bogs are present in this bog (Hansen, 1941c). Its sedimentary column in the area of sampling is 3.75 meters thick, and a thick layer of silt and volcanic ash occurs about half way up in the profile. This inorganic stratum was deposited directly over fibrous peat, and, upon resumption of organic sedimentation, two decimeters of limnic peat were laid down before fibrous peat was again formed. Evidently hydrarch succession was interrupted by the deposition of the silt. The volcanic glass is presumed to be equivalent to that in the Onion Flats profile and the pumice stratum in the Lake Labish sedimentary columns. The occurrence of the glass and silt may be merely coincidental or it may reflect increased erosion following the volcanic activity.

A second sedimentary column composed almost entirely of *Sphagnum* was obtained from a bog a few miles east of Scotts Mills in the foothills of the Cascade Range. The depth in the area of sampling is 8.4 meters, and about 4.5 meters is composed of almost pure *Sphagnum*. The bog has apparently been formed in a tributary dammed near its mouth by landslides, so the age is uncertain. Since there is no layer of volcanic glass present, the bog was probably out of reach of the source of the pumice interred in profiles farther west in the valley proper. This is the most typical Sphagnum bog noted south of the Puget Sound region and inland from the ocean, and it reflects the influence of the moister climate of the Cascade foothills. Many typical bog plants are present.

NORTHERN WASHINGTON AND IDAHO

The sedimentary columns from bogs and swamps in northern Washington and Idaho are distributed over an area that is not homogeneous with respect to climate and vegetation. Their sites range in altitude from 1,880 to 4,200 feet. In general conditions in this region are not so favorable for organic sedimentation as in the Puget Lowland. This is probably due

to the lower precipitation, the dry summers, and the shorter growing season over most of the region. The organic deposits have accumulated chiefly in basins with no outward surface drainage, so a seasonally fluctuating water table has not been important in limiting the rate of organic sedimentation. Sphagnum bogs are seemingly rare in this region, and the few that were observed do not support nearly so many bog plants as those in the Puget Sound region and the Pacific Coast. The other sites of maturing hydrarch succession support largely sedges, at present are in a sedge-meadow stage, and can be classified as sedge bogs. A bog at Fish Lake, near Lake Wenatchee, Washington, is at present in a sedge stage with some *Sphagnum* associated with it (Hansen, 1941f). All except one of the sites of hydrarch succession still retain a body of open water. In most cases the submerged and floating seres consist of typical hydrophytes. In two instances, however, a submerged stage composed chiefly of stonewort (*Chara* sp.) occupies the greater part of the standing water. Stonewort has not been observed by the author elsewhere in the Pacific Northwest as the most abundant species in the submerged sere. The mature swamps and sedge bogs in northern Washington and Idaho are usually invaded by species of willow and alder rather than by the bog shrubs common to the Puget Sound region and the Pacific Coast.

The average depth of eight pollen-bearing sedimentary columns in this region is about 6.5 meters, ranging from 1.25 to 12 meters (table 4). An average of about 2.5 meters or 38 per cent is composed of limnic peat, while an average of approximately 3.4 meters, or 52 per cent, consists of fibrous peat. In the two profiles mentioned above, Chara marl constitutes a substantial proportion of the limnic peat. The fibrous peat in most profiles is well preserved and raw, with no muck horizons to indicate lowering of the water table for any sustained period. The average thick-

---

TABLE 4

Sedimentary Columns from Northern Washington and Idaho

Location	Depth in m.[a]	Thickness overlying volcanic ash	Silt	Clay	Limnic peat	Fibrous peat
Fish Lake.....	8.5	3.5	2.0	0.25	3.25	3.0
Wenatchee.....	1.25	1.0			0.25	1.0
Bonaparte Lake	3.25	1.75			1.4	1.85
Priest Lake....	12.0	9.0			3.0	9.0
Bonners Ferry.	7.5	5.75	0.5		4.0	2.5
Newman Lake.	7.3	4.5			2.4	4.6
Liberty Lake...	7.0	4.5	1.0	0.5	3.5	2.0
Eloika Lake....	5.0	1.6	0.25		2.0	2.75
Average.....	6.5	3.9			2.5	3.4

[a] All figures in meters.

---

[1] All pumice attributed to Mount Mazama has been determined by I. S. Allison, Oregon State College.

ness of the sediments overlying the volcanic glass is 3.9 meters, with a range of 1 to 9 meters. As the ash is probably from the same source and of synchronous deposition in the several profiles, this means that the rate of organic sedimentation after the ash fall was nine times faster in the deepest than in the shallowest profiles. The thickest measured volcanic glass strata occur in the sedimentary columns of this region. The maximum found at Newman Lake, near Spokane, Washington, is about 0.25 meters thick, while a profile near Bonners Ferry in northern Idaho contains a 2-decimeter layer (Hansen, 1934e). Some sites lying even nearer the source of the ash, presumably Glacier Peak in north central. Washington, have thinner strata. This suggests that where the ash is thicker at distant points it was carried into the sedimentary basin by inflowing streams.

CHANNELED SCABLANDS OF THE COLUMBIA BASIN

Only four sedimentary columns of pollen-bearing sediments have been obtained from the scabland area of eastern Washington, so conclusions cannot be so definite as those based on a larger number. Apparently the two chief factors limiting hydrarch succession and organic sedimentation have been the semi-arid climate and the alkalinity of the substratum. Much of the permanent standing water, especially that farthest down-valley in the coulees, is too alkaline to support hydrarch succession to any great extent, as it is even more saline than salt marshes along the ocean (fig. 13). As previously mentioned, many of the scabland lakes lie in deep depressions excavated in the coulee floors, and are rock-bound. The shores are precipitous and the water too deep for rooted aquatics to gain a foothold except in some places at the ends of the lakes where sediments from incoming streams have provided a more gradually sloping bottom. In this type of lake hydrarch succession has been limited largely to early limnic-forming submerged stages.

In general the organic sediments in the Channeled Scablands contain a higher inorganic fraction than

TABLE 5

SEDIMENTARY COLUMNS FROM THE CHANNELED SCABLANDS

Location	Depth in m.[a]	Thickness overlying volcanic ash	Silt	Clay	Limnic peat	Fibrous peat
Cheney.......	8.8	6.2		0.2	5.6	3.0
Wilbur........	2.6	1.2	0.1		2.0	0.5
Harrington....	6.5	3.4	0.1		3.8	2.6
Crab Lake.....	6.75	5.0	0.5		2.75	3.5
Average.....	6.18	4.0			3.5	2.4

[a] All figures in meters.

those of other regions of this study. Some of the sediments are undoubtedly wind-borne. They lie directly upon bedrock, and little or no sand is present (table 5). The presence of a half-meter of pollen-bearing silt underlying the limnic peat in the Crab Lake profile indicates the high degree of erosion by incoming streams during the early stages of hydrarch succession. The proportion of mineral matter decreases upward in the limnic stratum, while the fibrous peat is almost entirely devoid of mineral matter except in the upper levels. The upper half meter is mucky owing to increased deposition of silt caused by more erosion, more seasonal flooding, and greater fluctuation in the water table since cultivation by man. This has resulted in subsidence due to oxidation and decomposition, especially at Crab Lake which has been drained and in cultivation for some years. Grazing has also been instrumental in causing subsidence and increase in the proportion of mineral matter in the upper levels. In a sedimentary column near Wilbur, Washington, a layer of almost pure diatomite is present, while the lower levels of a profile near Cheney, Washington, (Hansen, 1943d) are composed of Gastropod marl.

In spite of the semi-arid climate, the average depth of the pollen-bearing sedimentary columns in the scabland area is about 6.2 meters. A higher proportion of mineral matter than in the profiles of other regions has undoubtedly been partly responsible for this appreciable depth. Another reason for the unusual thickness in a dry region is the possibly greater age of these profiles than of others in the Pacific Northwest. They range in depth from 2.6 to 8.8 meters and the presence of the volcanic glass horizon at about the same relative stratigraphic position in all attests to their contemporaneity of accumulation. Many swamps in the Channeled Scablands were examined for possible pollen analysis, but most of them are too shallow and mucky and contain little or no pollen. Those used in this study are probably not typical as far as depth is concerned. An average thickness of about 3.5 meters, or over 56 per cent, is composed of limnic peat, while about 2.4 meters, or 38 per cent, consist of fibrous peat. The proportions of limnic and fibrous peat are the greatest and least, respectively, among all the sedimentary columns that rest upon glacial drift or its chronological equivalent. They suggest a longer eutrophic submerged sere before littoral vegetation became the chief contributor to the accruing sediments. The average thickness of sediments overlying the volcanic glass stratum is 4 meters.

CASCADE RANGE OF OREGON AND WASHINGTON

Fourteen pollen-bearing sedimentary columns have been obtained from bogs and swamps in the Cascade Mountains of Oregon and Washington and the Blue

TABLE 6

SEDIMENTARY COLUMNS FROM THE BLUE MOUNTAINS
AND CASCADE RANGE

Location	Elevation	Depth in m.	Sand	Silt	Clay	Limnic peat	Fibrous peat
Anthony Lakes 1	7000	3.0		0.2		2.4	0.4
Anthony Lakes 2	7000	4.0		0.1		3.3	0.4
Kachess Lake...	2200	4.2		0.2	0.4	1.2	2.4
Cayuse Meadows 1........	3700	2.4		0.1		0.9	1.4
Cayuse Meadows 2........	3700	2.8		0.2		1.2	1.4
Clear Lake.....	3496	2.4		0.2		1.1	1.1
Clackamas Lake	3340	2.0		0.2		0.3	1.5
Bend [b]........	5240	7.0	0.25			3.75	3.0
Mud Lake [c].....	5000	1.8	0.1				1.7
Willamette Pass.	2900	2.75				1.5	1.25
Big Marsh [a].....	4000	1.8	0.1			0.3	1.4
Diamond Lake [a].	5196	0.5		0.1		0.1	0.3
Munson Valley [a].	6200	2.0				0.4	1.6
Rogue River [a]...	4000	4.0	0.8	0.4		0.6	2.2
Prospect [a]......	2500	3.2		0.2		0.2	2.8

[a] Underlain with Mount Mazama pumice.
[b] Mount Mazama pumice at 4.5 m.
[c] Underlain with Devil's Hill pumice.

Mountains of northeastern Oregon. The sites of these pollen-bearing sediments range in altitude from 2,500 to 7,000 feet, and as a group they are spoken of as montane. Some of these sites of hydrarch succession lie on the east slope, some on the west slope, while others lie practically astride the crest of the Cascade Range. As a group they show some disparity as to possible age, and the adjacent area within range of pollen dispersal to each site varies considerably with respect to climate and vegetation. At least five of these profiles rest directly upon Mount Mazama pumice and may be assumed to be of contemporaneous accumulation, while another located at Mud Lake rests upon pumice from a different source,[2] although of postglacial origin (Williams, 1942).

The average depth of all montane profiles is 2.8 meters with a range of 0.5 to 7.0 meters (table 6). The average depth of those resting upon pumice is 2.3 meters, while for those not on pumice it is about 2.8 meters. These are the lowest average thicknesses of the several groups of sedimentary columns. The comparative shallowness of these sedimentary beds may be due to the short growing season, the dry summers, and the coldness of the water and substratum. In addition to these unfavorable conditions, those resting upon Mount Mazama pumice are younger than bogs resting upon glacial drift, as the eruption of Mount Mazama occurred sometime after the maximum of continental glaciation (Williams, 1942; Allison, 1945). Recession of Pleistocene valley glaciers probably occurred later than that of the continental ice but, nevertheless, deglaciation had pro-

[2] Determined by I. S. Allison.

ceeded to an elevation of more than 5,000 feet in at least one valley on the east slope of the Cascades before the eruption. In a bog near Bend, Oregon, that lies in this glaciated valley, two layers of pumice are present. In a 7-meter profile one stratum occurs at 4.5 meters and the other at 2 meters. In a previous paper (Hansen, 1942a) the upper stratum was assigned to the eruption of Mount Mazama owing to the thickness of the overlying sediments as compared to those of bogs that rest directly upon the pumice. A petrographic examination of the pumice discloses, however, that the lower stratum came from the main eruption of Mount Mazama (Allison, 1945), while that at 2 meters was derived from some unknown source. It may represent an eruption of Devil's Hill, located about 10 miles west, which is known to have been active since deglaciation (Williams, 1944). Its output of pumice, however, was small and of only local distribution. Bend is located in an area where the Mount Mazama pumice is from 1 to 3 decimeters thick (Williams, 1942). The presence of 4.5 meters of organic sediments above the pumice in the Bend profile represents the maximum amount of organic deposition in this area since the eruption of Mount Mazama. The deposition of several to many feet of pumice must have terminated hydrarch succession in many basins. In some it may have resumed afterwards, while in others changes in topography and drainage prohibited subsequent organic sedimentation. In only one other profile of this study was sedimentation resumed after the deposition of the pumice. In a 2.75-meter column on Willamette Pass, Oregon, just west of the Cascade divide, a layer of pumice of Mount Mazama origin (Hansen, 1942c) occurs at 2.5 meters.

The sedimentary columns that rest on Mount Mazama pumice including those with an interbedded stratum are composed of an average of 0.6 and 1.8 meters of limnic and fibrous peat respectively, while the other montane profiles consist of an average of 1.5 meters of limnic and only 1 meter of fibrous peat. This difference in the thickness of limnic peat suggests that the pumice fall destroyed much of the biota in the region, and the lakes remained strongly oligotrophic until littoral vegetation developed dystrophy which has persisted to the present. The fibrous peat of the montane bogs is composed largely of sedge and sedge-like plants, but three of the columns contain considerable *Sphagnum*. One near Bend contains considerable Hypnum peat, while one at Mud Lake is composed of an undetermined species of moss in addition to sedge. The fibrous peat in the montane bogs shows little or no oxidation, decomposition, and compaction. Some of the sites of hydrarch succession support bog plants, but not nearly so many as those along the Coast and in the Puget Sound region.

Two peat columns from a bog surrounding a lake in the Blue Mountains of northeastern Oregon rest upon

Bureau of Reclamation photo, Coulee Dam, Washington

FIG. 13. Soap Lake in Lower Grand Coulee, Washington. Far downstream in the Grand Coulee the water is so highly saline as to eliminate all visible signs of life.

Bureau of Reclamation photo, Coulee Dam, Washington

FIG. 14. Sand dunes in the Potholes area of eastern Washington.

Photo by L. S. Cressman

FIG. 15. View of a typical tule swamp in south central Oregon. Lower Klamath Lake was like this before drainage in 1917.

Photo by L. S. Cressman

FIG. 16. Bed of Lower Klamath Lake after drainage followed by erosion and fire. Upper surface of peat in foreground represents approximately the original surface after drainage.

volcanic glass (Hansen, 1943a), A layer more than 2 decimeters in thickness is present, and it was impenetrable to a greater depth. The source of the ash is not known but it hardly seems possible that it records the same volcanic activity responsible for the ash stratum in most of the Washington pollen-bearing profiles. However, its thickness in the profile may have been increased by water transport from the surrounding hydrographic basin. Two peat profiles from a bog near Mount Adams, Washington, are underlain with pumice, and also contain a stratum of volcanic ash at higher levels (Hansen, 1942e). The proximity of Mount Adams, Mount Saint Helens, Mount Hood, and Mount Rainier, all volcanic cones, makes it imprudent to assign a source to this ash without further study. Mount Saint Helens was active in 1802 and again as late as 1842 (Lawrence, 1938, 1941), and offers a possible source for the ash. It is also possible that it represents the same eruption as the ash stratum common to most of the Washington pollen-bearing profiles.

In a 6-meter sedimentary column from Farger Lake, about twenty miles north of Vancouver, Washington, a layer of water-washed pumice occurs at 1.1 meters. It is of unknown source and was undoubtedly carried into the lake by streams heading in the mountains to the east. A bog at Silver Lake, near Kelso, Washington, 3 meters deep, is underlain with volcanic glass, also of unknown origin. The proximity of Mount Saint Helens, however, with streams heading in its vicinity, suggests this volcano as the source of origin.

A montane bog near Lake Kachess on the west slope of the Cascades in central Washington is not included in the montane group because of its isolation from the sites of the others, mostly in Oregon. In a 4.2-meter sedimentary column a stratum of volcanic ash is present at 3.2 meters. This ash is probably equivalent to that common to most of the Washington peat profiles. Its age is probably similar to, or only slightly younger than, the bogs of the Puget Sound region. Of interest, but not particularly pertinent to this study, is the occurrence of an Equisetum swamp noted at high elevation in the Mount Adams region. A swampy area comprising about 25 acres, and composed of almost a pure stand of *Equisetum fluviatile*, is underlain with about 1 meter of silica derived from the decomposition of the scouring rush. An abundance of silica available for assimilation by this plant is provided by the volcanic glass and pumice in this region.

NORTHERN GREAT BASIN

Seven pollen-bearing sedimentary columns were obtained from the northern Great Basin of south central Oregon. As mentioned above, the lakes of the northern Great Basin probably date back at least to late Pleistocene, and their levels have fluctuated correlatively with the continental and mountain glacial

stages, and during the postglacial with climatic cycles. The pollen-bearing sediments in these basins, therefore, may be as old as those that rest upon glacial drift or its chronological equivalent. The rate of accumulation has been much slower than in the Puget Lowland because of the less favorable environmental conditions. During the warm, dry period between 8,000 and 4,000 years ago organic sedimentation probably was retarded, and on some sites may have ceased entirely, resulting in unconformities of unknown duration. This hiatus may have been increased by removal of sediments by deflation if the water table sank several feet below the surface.

Four sections were obtained from Lower Klamath Lake. This extensive shallow lake was drained in 1917, and subsequent cultivation has resulted in considerable subsidence as well as loss of the upper fibrous peat layers owing to burning and deflation (fig. 16). Sections were obtained where most of the original organic sediments were intact. Silt and volcanic glass are present throughout the sections, the latter indicating water and wind transport of volcanic material from the Cascade Range to the northwest. In the southern part of Lower Klamath Lake bed the removal of several feet of fibrous peat and a lesser thickness of limnic peat has exposed many artifacts occurring along a beach line of a prehistoric level of the lake. These artifacts were apparently left by early man toward the close of the warm, dry middle postglacial when the lake bed had already begun to be reinundated (Cressman, 1942). Complete inundation of the lake bed and subsequent hydrarch succession formed several feet of peat, much of which has been removed by wind and fire since drainage. The occurrence of the artifacts and their implications mean that an unconformity is present in the pollen profiles of these sections. This time gap may be magnified by the loss of sediments by fire and deflation during the previous exposure of the lake bed.

A fifth section was obtained from Klamath Marsh located about 15 miles due east of Crater Lake. This marsh rests upon Mount Mazama pumice, which is about 10 feet thick in this region (map 2), and an excellent record of post-Mount Mazama forest succession in the region is recorded in the column. A sedimentary column was obtained from the Warner Lake basin, located about 110 miles east-southeast from Crater Lake. A layer of Mount Mazama pumice occurs in this profile which serves to correlate the pollen profiles with those farther west and north. A seventh column was obtained from Chewaucan Marsh, located about 80 miles east-southeast from Crater Lake. The Chewaucan Marsh basin was covered by Pleistocene lakes when they were high enough to extend through the gap at the southeast corner of Summer Lake basin (Allison, 1945). This section likewise contains a Mount Mazama pumice horizon.

TABLE 7

SEDIMENTARY COLUMNS FROM THE NORTHERN GREAT BASIN

Location	Depth in m.[a]	Thickness overlying pumice	Silt	Limnic peat	Fibrous peat
Laird's Bay.........	2.7			1.8	0.9
Narrows No. 1.......	3.0			2.0	1.0
Narrows No. 2.......	3.3			2.4	0.9
Klamath Falls.......	2.5		0.1	2.4	
Klamath Marsh......	2.5	2.5	0.1[b]	1.9	0.5
Chewaucan Marsh....	2.4	1.2	0.4	1.7	0.3
Warner Lake basin....	3.5	2.6	0.1	2.2	1.2
Average..........	2.9			2.1	0.7

[a] All figures in meters.   [b] Pumice.

The average depth of the six sections is 2.9 meters, ranging from 2.5 meters to 3.5 meters (table 7). An average of 2.1 meters or 72 per cent is composed of limnic peat and about 0.7 meter or 28 per cent consists of fibrous peat. The average depth to the pumice level in the three sections is about 2.2 meters, about the same as for those in the Oregon Cascades.

SUMMARY

The wide range of climate in the Pacific Northwest has resulted in many kinds of pollen-bearing sediments from the standpoint of composition. Although many species of organisms have contributed their remains to the organic sediments of this study, structurally they are classified into limnic and fibrous peat. There are several kinds of underlying sediments owing to the several methods of origin of the depressions in which they have accumulated. These include glacial drift, dune sand, bedrock, volcanic materials, silt, clay, and sand. Some of the underlying sediments have sufficient pollen grains to extend the forest history back a short time beyond the beginning of organic sedimentation. Most of the lakes in which the sediments have accumulated were of the oligotrophic type at the beginning, especially those in the glaciated region. Eutrophication was soon attained, however, as hydrarch succession progressed, and in turn some of the lakes were converted to the dystrophic type in which brown peat was formed. No attempt has been made to interpret the typological succession of the sedimentary column into terms of extra-aquatic environment, as has been done in northern Europe.

Optimum conditions for peat formation have existed in the Puget Sound region and along the Pacific Coast. The most unfavorable conditions have existed in the Oregon Cascades, where the short growing season and the dry summers have limited the thickness of organic sediments. In eastern Washington and the northern Great Basin of south central Oregon development of alkalinity has slowed the rate of organic sedimenta-

tion, and at some sites high salinity has entirely prohibited hydrarch succession. In some swamps in this area the surface has become sufficiently alkaline so as to support halophytic marsh plants. Most of the Sphagnum bogs are found in the Puget Sound region and along the Pacific Coast where the summers are cool and the winters relatively warm and moist. Usually the *Sphagnum* is underlain with sedge which in turn is underlain with fine limnic or sedimentary peat, derived from fine detritus that settled to the bottom of the open lake. The deepest bogs of this study are located in the Puget Sound region, two of them attaining a depth of 13 meters. The average depth of 16 bogs in this area is almost 8 meters. In eastern Washington and Oregon, Sphagnum bogs are absent because of the dry summers and cold winters and alkalinity of the substratum. Many tule swamps have formed appreciable depths of fibrous, mucky peat, while sedges and various sedge-like plants have been the chief contributors on other sites. In Washington, about thirty of the pollen-bearing sedimentary columns contain a layer of volcanic ash which probably had its source from volcanic activity in the Glacier Peak region of north central Washington. This layer is of extremely great value as a common time marker, and it has made much easier the task of correlation of forest succession and climatic trends between the several areas. It also has made the chronology far more accurate than would otherwise have been possible.

In the coastal strip of Oregon and Washington the peat sections are not very thick, probably because of the youthfulness of the depressions. Only a few may be early postglacial in time of origin. Most of the bogs are of the Sphagnum type which originated in oligotrophic lakes which were converted into dystrophic lakes with no intermediate eutrophic stage. The distance from the source of the Washington volcanic ash has prohibited its reaching the site of the coastal bogs.

Postglacial organic sediments are not common in the Willamette Valley because of the dearth of undrained depressions and the dry summers. The sediments are largely of swamp origin, and fluctuation of the water table has resulted in mucky, partially oxidized peat. In the lower part of the valley a layer of pumice occurs in the upper third of the sections. The source of this pumice is unknown, but Mount Saint Helens in southern Washington seems to be the most logical source. Only two Sphagnum bogs were found in the Willamette Valley, and these lie in the foothills of the Cascade and Coast Ranges where precipitation is greater than in the valley proper.

Although northern Washington and Idaho were glaciated, peat bogs are not common, perhaps because of the well-drained terrain. The low summer precipitation and the cold winters have limited the amount of postglacial organic sedimentation. Two of the bogs sampled contained a thick stratum of highly calcareous stonewort marl which is not common in the Pacific Northwest. In this area the strata of volcanic ash are thickest, owing probably to the strong westerly winds which distributed it directly east from Glacier Peak, its apparent source.

The aridity of the Channeled Scablands of eastern Washington is not favorable for organic sedimentation, but four sections were obtained from this region. All sections lie directly upon bedrock, in coulees that were cut or deepened by glacial meltwater that spilled over the Spokane River divide during retreat of the last glacier and probably earlier ice sheets. The dry summers causing fluctuation of water tables and the income of alkaline soil from the basaltic bedrock have retarded or prohibited hydrarch succession, but the average depth of four sections is over 6 meters. Pollen grains are abundant at most levels and the pollen profiles offer significant data for the interpretation of postglacial climate.

The Cascade Range of Oregon and Washington abounds with peat bogs, but their inaccessibility has prevented the analysis of as many columns as desired. As a group they are the shallowest deposits of this study. In Oregon most of the sections overlie pumice from one of the several postglacially active volcanoes, while others contain one or more pumice strata. In Washington the common volcanic ash layer is present. The several different ages of the sections, as indicated by the pumice and underlying glacial drift, provide valuable chronological data which aid in the dating of the eruption of Mount Mazama.

The closed basins of south central Oregon have contributed several important sedimentary columns to this study. In addition to providing excellent pollen profiles for the interpretation of vegetational and climatic cycles, the occurrence of pumice strata and artifacts lends support to Pacific Northwest postglacial chronology by offering a correlation with that of the Great Basin. Some sections may be as old as those that rest upon glacial drift, while others overlie Mount Mazama pumice and are much younger.

GENERAL LATE–GLACIAL AND POSTGLACIAL CHRONOLOGY

No adequate basis as yet has been formulated for determining in terms of years the periods of time involved in the several glacial and interglacial stages. Data on postglacial chronology in the Pacific Northwest are somewhat limited by the lack of such intensive investigations as have been carried on in other parts of the continent. Whereas it can be generally assumed that the glacial stages in the Pacific Northwest were chronologically equivalent to those west of Hudson Bay and the Mississippi River, progressive shifting of storm tracks as the size and extent of the

glaciers grew, and retracing of the storms tracks as the glacier wasted probably resulted in time differentials of unknown and immeasurable magnitude. Further deviation from synchroneity was caused by the differences in topography, direction, and distance from centers of accumulation, and in climate at the outermost termini of the several lobes.

Considerable work has been done on the chronology and correlation of the late-glacial and postglacial time in eastern North America and in Europe, and various estimates have been made of their duration. These estimates are based upon varved clay studies, weathering and leaching of till sheets, recession of Niagara Falls and other falls, development of shorelines, position of ancient shorelines, rate of deposition and thickness of organic and inorganic sediments, erosion by rivers, growth rings in trees, rate of accumulation of salts and degree of salinity in Great Basin lakes, archaeologic relics, oscillations of mountain glaciers, rate of biotic migrations, climatic cycles interpreted from pollen profiles, and other biologic, cosmic, and geologic processes. Some of these estimates involve only a part of late-glacial and postglacial time, while others may include all or most of it. One of the major problems is to assemble in correct sequence the chronological segments gleaned from the several sources as well as to correlate the parts or whole from one study with those from others. It is remarkable how some of these chronologic segments fit together or correlate with others to form an integrated and continuous time table for the trends and events of postglacial time. The figures arrived at have varied considerably over a period of years, but continual adjustment and compromise in view of the ever accumulating chronological data have tended gradually to narrow this range of variation. Much of the disparity in chronological consideration has been caused by independent studies made in different localities, and conclusions drawn without the compromising effect of other data. The chronologist of today realizes the importance of evaluating the data from all sources, fitting them together in sequence, and correlating the parts.

As the amount of chronological evidence continues to accumulate, the general tendency has been to reduce the time estimates since the waning of the last stage of the Wisconsin glaciation. These estimates vary from 80,000 years since the Wisconsin maximum (Croll, 1876) in parts of North America, to 8,500 years since the ice had practically disappeared from parts of Europe (De Geer, 1910). Some workers refer to the Postglacial as the time since the last glacial stage reached its maximum and began its final retreat. Others delimit postglacial time as the period since ice-wastage uncovered a given region of its ice. The first method provides a more nearly common time-scale for a much broader region than the latter because local conditions at the perimeter of the ice sheet must have had considerable influence upon the time and rate of recession. In considering the age of sedimentary column a third factor arises, namely the time lapse between deglaciation and the beginning of sedimentation. It is important in this study to distinguish in relative figures if possible the time differential involved in the initiation of pollen-bearing sedimentation, not only in the several physiographic provinces, but also within smaller areas more nearly uniform with respect to postglacial trends.

In view of the fact that ice sheets last retreated from central Illinois, southern Ohio, and Indiana some 40,000 years ago, Connecticut some 30,000 years, northern Washington 25,000 years, and Stockholm 10,000 years ago, and that the final disappearance of the glaciers in Canada and Scandinavia occurred about 5,000 years ago, it would eliminate much confusion if delimitation of the Postglacial were based on a common factor in all regions. If each worker in geochronology would clearly define what he means by "postglacial" time, and correlate it with other postglacial chronology, a better understanding of the entire problem would be had.

The Wisconsin glaciation included the western (Cordilleran and Keewatin) and the eastern (Labrador and Patrician) ice, the line of demarcation extending along western Hudson Bay and the Mississippi River (Antevs, 1946). The western ice (Keewatin) is believed to have had an early maximum, the Iowan, some 65,000 years ago, and a later maximum, the Mankato, some 25,000 years ago. The eastern ice had two intervening maxima, and a final maximum, the Valders, contemporaneous with the Mankato, 25,000 years ago (Thwaites, 1943). In the Puget Sound region, the last glaciation, the Vashon, is probably equivalent to the last maximum of the Wisconsin, the Mankato. In northern Washington and Idaho, the final glacial stages, the Okanogan and the Spokane lobes, were probably contemporaneous with the "Wisconsin" (Flint, 1937). In at least two areas in the Cascades of Washington the last stage of mountain glaciation is thought to have been concurrent with the last Wisconsin (Mackin, 1941; Page, 1939). The last stage of glaciation in the Sierra Nevada, the Tioga, is also believed to have been synchronous with the last Wisconsin (Blackwelder, 1931). In general, then, it can be said that glacial geologists and other workers in glacial chronology are in substantial agreement as to the above correlations of the last glacial stage in various parts of North America.

The time estimates mentioned above for the maxima of the Iowan and Mankato are not mere guesses but are based upon various geologic and astronomic data. A temperature minimum about 72,000 years ago is suggested by the Spitaler-Milankovich-Koppen glacial hypothesis and chronology, while the degree of weathering and leaching of Iowan drift in Iowa suggests exposure for a period of about

55,000 years (Kay, 1931). Antevs (1941) has taken the approximate mean of these figures, or 65,000 years, as the age of the Iowan maximum. The age of 25,000 years for the Mankato glacial maximum is based on the varved clay chronology and correlation of the ice border at St. Johnsbury in Vermont, via the Port Huron morainic system and other correlative moraines, with the ice border of the Mankato culmination west of the Mississippi River (Antevs, 1945). According to the astronomic chronology there was a temperature minimum about 22,000 years ago. Calculations based on the rate of retreat of St. Anthony Falls suggest that the Keewatin ice persisted near the northeast limits of the glacial Lake Agassiz as recently as 9,000 years ago (Leverett, 1932). The latest estimate places the age of the falls at 20,000 years (Thwaites, 1937). Thwaites (1937) sets a figure of 50,000 years since the maximum of Wisconsin glacial stage in the Mississippi Valley, and Thornbury (1940) estimates that the oldest Wisconsin drift in Indiana has been undergoing leaching for 45,000 years. In each region the culmination of the ice sheets was soon followed by retreat.

The rate of retreat varied in different regions at different times. De Geer (1940) found by varve counts that about 6,000 years were required for the ice in southern and central Sweden to retreat a distance of about 575 miles. The beginning of the retreat may have occurred approximately 14,000 years ago. The time since complete wastage of the ice in Sweden is not known, but it is perhaps some 5,000 years. Antevs (1922, 1928) has found that about 9,500 years were involved in the retreat of the Labrador ice from the Long Island moraines, which are excluded, to the White Mountains, a distance of 230 miles. This results in an average of 41 years per mile, while the rate between Hartford and St. Johnsbury was 22 years to the mile. As the distance from Long Island to the center of accumulation in Labrador is about 950 miles, about 30,000 years were required for complete dissipation of the Labrador ice if the mean of these two rates is applied.

De Geer (1910) suggested that the bisection of the ice remnant west of Ragunda in northern Sweden be taken as the transition between the glacial and the postglacial ages. This was dated as 6840 B.C. by Lidén (1938), which makes the Postglacial about 8,800 years. This is considered as the length of the Postglacial by Swedish and Finnish geologists, but it can be applied to only a limited region. Antevs (1931) suggested using the temperature, especially the summer temperature, as a basis for defining postglacial time. He sets the beginning when the temperature, rising from the time of the last glaciation, in the southern parts of the deglaciated area had risen to that of the present time. On the basis of the warm summer cycle from 15,000 to 4,500 years ago, this may have been about 9,000 years ago, which correlates with the Swedish varved clay chronology. · Farther south the temperature rise was more advanced, and Antevs (1941) has set the beginning of the Postpluvial at 10,000 years ago in the Great Basin.

The pluvial stages of the Southwest as recorded by the fluctuations of large lakes that are now dry, or nearly so, probably reflect the maxima of the continental glacial stages in northwestern North America. Antevs (1941a, 1946) has interpreted the history of Lake Bonneville and Lake Lahontan in the Great Basin as correlative with the climatic trends during the Wisconsin glacial period. Lake Bonneville and Lake Lahontan proper, as recorded by the highest shorelines, may represent the pluvial stage in that region which corresponded to the Iowan continental glaciation. Lowering of the lakes occurred between the Iowan (early-Wisconsin) and Mankato maxima. A second but lower shoreline is thought to be correlative with the Mankato maximum. The lakes that formed these shorelines are named Lake Provo in the Bonneville Basin and Dendritic Lake in the Lahontan Basin. The two pluvial stages that were probably the chronological equivalents of the Iowan and Mankato maxima are called the Bonneville and Provo Pluvials respectively. These pluvials also probably correspond to the Tahoe and Tioga glacial stages in the Sierra Nevada.

Antevs (1945) believes the last glaciation culminated in the Pacific Northwest 25,000 years ago. The distance from the southern terminus of the Puget Sound glacier near the Chehalis River to the Canadian boundary is not over 200 miles. If we assume a rate of retreat of about 25 years to the mile, about 5,000 years would have been required for the ice to recede from the Puget Sound region of western Washington. If organic sedimentation began soon after the potential sites were freed of ice, then the oldest sedimentary columns may be almost 25,000 years old. According to Antevs' chronology the first 15,000 years represent the late-glacial, and only the last 10,000 years the Postglacial. Sedimentary columns on glacial drift in the Oregon Cascades and in northern Washington east of the Cascades and in northern Idaho are probably younger with a possible maximum age of not more than 15,000 years. Those sedimentary columns situated beyond the limits of Pleistocene glaciation are probably not older than those on glaciated sites because of the close relationship of the physiographic history of these regions to glaciation. The late-glacial and postglacial trends of climate and forest succession as interpreted from the pollen profiles, the thickness of the sedimentary columns, the stratigraphic position of the pumice and volcanic ash strata, and the typological succession of the sediments, correlated with other chronological evidence from studies in archaeology, geology, vulcanology, and glaciology in the Pacific Northwest and the Great Basin suggest that an average of 18,000 years for the age of those

sedimentary columns resting upon glacial drift or its chronological equivalent is reasonable.

The foregoing discussion then defines the chronology of late-glacial and postglacial time as applied to Sweden, eastern North America, the upper Mississippi Valley, Ohio and Indiana, the Sierra Nevada, the Great Basin, and the Pacific Northwest. For the sake of brevity all of the time since the maximum of the late-Wisconsin glaciation will be referred to as merely postglacial without differentiating the proportion of late-glacial and postglacial time except where it is of special chronological significance.

The layers of volcanic ash and pumice in or underlying the peat beds are also of chronological value, as they mark distinct volcanic eruptions and serve to separate time into smaller units. The greater the number of such units into which late-glacial and postglacial time can be divided, the more accurate the dating of the specific events and trends becomes.

The correlation of Spitaler's warm and cool summer cycle hypothesis with glacial and postglacial chronology tends to substantiate the general estimates given above. In the following table Antevs (1933) has correlated the periods of warm and cool summers from Spitaler's data with postglacial climate and chronology in Sweden and with the present knowledge of the ice waning in North America.

Warm Summers	Cool Summers
Since 4,500	Late Postglacial
years ago	
15,000 to 4,500	⌈ Middle Postglacial
years ago	⟨ Early Postglacial
	⌊ Younger Late-glacial
25,000 to 15,000	Middle Late-glacial
years ago	
35,000 to 25,000	Early Late-glacial

The Pluvial and post-Pluvial chronology for the Southwest is shown in the following table (Antevs, 1941, 1946). .

	⌈ Late (since 2000 B.C.)	Especially in the first half moderately moist.
Postpluvial	⟨ Middle (6000–2000 B.C.)	Warmer and drier than at present
	⌊ Early (8000–6000 B.C.)	Climate at beginning as today, but getting warmer and drier.
Provo Pluvial (23,000–10,000 years ago).		Decidely cooler and moister than at present, with the culmination 25,000 to 22,000 years ago.

RATE OF ORGANIC DEPOSITION

In a preceding chapter an estimate was made of postglacial time as based upon geologic and astronomic data from several sources. Consideration of the various estimates and data involved in reaching them has resulted in the writer's own general estimate of an average of about 18,000 years for the lapse of time since organic sedimentation had its beginning on sites occupied by the continental glacier in the Pacific Northwest. The thickness of organic sediments that have accumulated during postglacial time is of some value in supporting or refuting the estimates founded upon other data. The magnitude of the pollen-bearing sedimentary columns does not in itself provide a criterion for detailed postglacial chronology, but at least it tends to restrict the range of estimates within reasonable limits as deduced from the logical rates of organic sedimentation. The rate of peat deposition is perhaps less influenced by extra-aquatic conditions than that of inorganic sediments. Once organic sedimentation is under way in standing water, the rate for each type of sediment remains fairly constant. This becomes more so in those areas where the principal factors controlling the rate of deposition are not at a critical minimum. In the Puget Sound region the undrained basins with a more or less constant water table and small annual range of temperature provide sites for a more constant rate of organic sedimentation than in the Channeled Scablands of eastern Washington. In the latter area slight fluctuations in annual precipitation might greatly retard or increase the rate of deposition, whereas in the Puget Sound region small fluctuations in annual precipitation would have little influence in changing the rate of deposition. In using sedimentary columns in computing a time scale, the possible existence of unconformities must be considered. Sources of unconformities include fire, removing strata of fibrous peat; dry periods with cessation of deposition, oxidation and subsidence, and loss of surface sediments by deflation; and in rare cases removal of sediments by erosion. Unconformities from some of these causes occur in the sedimentary columns of this study, but they are immeasurable and can be used only in a general way to adjust the age of the sediments in which they occur after their general age has been computed on the basis of average rate of accumulation.

The more numerous the profiles from a homogeneous geomorphic, climatic, and floristic province upon which an average thickness can be based, the more significant the rate-time computations become. Absolute chronology in terms of years cannot be determined, however, because the rates of deposition must be based on an estimated chronology to begin with. The presence of the volcanic ash stratum in the Washington profiles is an excellent chronological marker and provides a definite comparative time scale for a part of the sedimentary column. The sediments overlying the ash stratum represent the same period of time for their deposition in all profiles regardless of their thickness. This is important because it reveals the great amount of variation in the rate of organic sedimentation that has occurred, not only in the several different climatic areas, but also within the same area. Without this correlating evidence, one might be tempted to assume that 1 meter of sediments represents only a small portion of the time required to accumulate 9 meters. Such a disparity in relative thickness actually occurs (table 1). The maturity of hydrarch succession in a given sedimentary basin when the ash fell is perhaps the chief factor determining the amount of subsequent organic sedimentation. The greater the maturity, the less the accumulation since the deposition of the ash. The ash stratum also serves to correlate chronologically similar trends of forest succession in widely separated areas, which on the basis of their relative stratigraphic positions in their respective sedimentary columns would seem to be far from synchronous.

In computing the rate of organic sedimentation upon the basis of estimated postglacial time, the average depth of only those sedimentary columns lying upon glacial drift or its chronological equivalent can be used. The age of those sediments that apparently had their origin at some unknown later time after deglaciation must be estimated by applying the rate of deposition as based upon the others of more definite age. In a certain sense, this seems to be going in circles, but the estimates thus attained may be better than none. The average thickness of thirty pollen-bearing sedimentary columns resting upon glacial drift or its chronological equivalent is 7.2 meters. This average includes the small and varying thickness of inorganic pollen-bearing sediments that underlie the organic. These limited strata are not segregated because their pollen profiles and their interpretation in terms of early postglacial vegetation history also contribute to the chronology. The rate of deposition, using 18,000 years for the duration of postglacial time, would then be about 2,500 years per meter. It seems feasible to adjust this figure in accordance with the obvious differential rate of deposition for limnic and fibrous peat. The limnic stratum, underlying the fibrous peat, is of slower accumulation as suggested by its composition, structure, and its greater compaction due to its stratigraphic position. Lesquereux (1885) says that peat at the bottom of a deposit may be compressed to less than one-eighth of its original volume. Sears and Janson (1933) believe that in the earlier deeper phases of postglacial lakes the process of silting was much slower than subsequent peat formation. It is presumed that the term silt is applied to sediments comparable to gyttja. The relative thickness of sediments overlying the volcanic glass stratum in the several profiles provides irrefutable evidence of the great variation in the rate of organic sedimentation.

The slower rate of limnic peat formation is further suggested by the relative abundance of pollen grains contained in it and in fibrous peat. The greatest concentration of pollen grains occurs in limnic peat, decreasing upward in the fibrous peat and downward in the coarser inorganic materials. This was also found to be true by Voss (1934) who records this evidence by means of a frequency curve adjacent to his pollen diagrams. As the average concentration of the pollen rain probably varies only slightly over a period of years, the relative concentration of pollen grains serves as an index to the relative rates of organic deposition, and perhaps to a rough time-scale gradient for the entire sedimentary column. The relative pollen concentration suggests about a 1 to 2 ratio in the rate of deposition between limnic and fibrous peat. The gradient of the pollen frequency upward and downward from its highest point suggests that the application of a logarithmic scale might be applied to gradients in rate of deposition. The significance of this is lost when the variation in depth of the several profiles is considered, and the results would hardly be commensurate with the work involved. An adjustment from 2,500 years per meter for the general rate of deposition to 3,500 years for a meter of limnic peat and 1,700 years for a meter of fibrous peat is therefore considered to be reasonable. The average depth of limnic peat in the 30 profiles is 3.4 meters and would represent about 12,000 years, while the average depth of the fibrous peat is 3.5 meters and would represent about 6,000 years. The small average thickness of the pollen-bearing mineral matter underlying the limnic peat represents a small additional amount of time of unknown magnitude. These silts and clays, and in a few cases sand, are undoubtedly of comparatively rapid deposition, as suggested by the low concentration of pollen grains interred. The application of these figures to individual profiles, as well as to the average thickness of the sediments for each region, reveals their general inadequacy. Sears and Janson (1933) have estimated upon the basis of measured laminations that one foot of peat in the Erie Basin requires 300 years for its formation. This figure refers to the fibrous peat near the surface, and the underlying, slower accumulating limnic peat would certainly require much longer. The impracticability of estimating rates of deposition for unit thicknesses of organic sediments and applying these estimates to sedimentary columns is further shown by the wide range of estimates by various authorities for the depositional time required for one foot of organic sediments. These estimates, by a score of scientists, range from 2 to 1,665 years for various parts of the Northern Hemisphere (Sears and Janson, 1933). In the Pacific Northwest the rate of organic sedimentation for the average of the profiles that rest upon

glacial drift or its chronological equivalent in each region varies from 2,400 to 2,900 years per meter. For individual profiles it has possibly varied from 625 to 6,000 years per meter.

The sedimentary columns from three different regions in the Pacific Northwest are located beyond the limits of Pleistocene continental glaciation, so this general time marker cannot be used as a criterion for their ages. It seems probable that most of them had their beginning some time after the glacier in the Puget Lowland had wasted and are younger than those resting upon glacial drift. The average depth of ten sedimentary columns on the Oregon and Washington Coast is 4 meters, of which 0.9 meter is limnic peat and 3.4 meters are fibrous peat. Applying the adjusted rates of deposition, these profiles would represent an average of about 7,200 years. The two deepest profiles, 11.75 and 7 meters, possibly had their origins in early postglacial time. Eliminating these two profiles would leave an average of about 2.7 meters, and reduce the age of the sand dune profiles to about 6,000 years, possibly a more reasonable estimate.

Five sections that rest upon Mount Mazama pumice and two with an interbedded stratum are composed of an average of 0.6 and 1.8 meters of limnic and fibrous peat respectively, which taken together upon the basis of the estimated rates of deposition would represent about 5,000 years. This estimate may represent the approximate time involved in the deposition of the peat, but it is undoubtedly too short for the total post-Mount Mazama time. The climatic eruption of Mount Mazama is estimated to have occurred between 4,000 and 7,000 years ago (Williams, 1942), but Allison (1945) has shown by correlating lacustrine records of the eruption with the chronology of postpluvial lake-level fluctuations in the northern Great Basin that it may have taken place between 10,000 and 14,000 years ago. The shallowness of the peat profiles resting upon Mount Mazama pumice may be explained in part by a long lapse of time after the volcanic activity before organic sedimentation began. The deposition of the thick pumice mantle probably destroyed all life in the area of its fall, and several thousands of years may have passed before peat-producing organisms were able to migrate to the lakes and ponds, because of their widely separated sites. If, on the other hand, organic sedimentation began soon after deposition of the pumice, and the sedimentary columns represent most of post-Mount Mazama time, then the rate of organic sedimentation was exceedingly slow, possibly because of unfavorable environment, and the average depositional-rate estimates cannot be applied with any degree of reliability. The occurrence of a dry period following the eruption of Mount Mazama may have retarded peat deposition in this area.

Six other montane sedimentary columns of the

Cascades and Blue Mountains are perhaps older than those resting upon pumice, and date back to the time when their sites were freed of Pleistocene mountain glaciers. Their average depth of 2.8 meters, with 1.5 and 1.0 meter of limnic and fibrous peat respectively, would require about 7,000 years for their deposition. Conditions somewhat removed from the optimum are probably responsible for a slower depositional rate than that upon which the estimates are based. As mountain deglaciation presumably lagged behind melting of the continental ice sheet, and cool conditions still prevailed at high altitudes, perhaps 13,000 to 15,000 years may be represented by these montane sedimentary columns.

The sedimentary profiles from Lower Klamath Lake represent about 8,000 years on the basis of the estimated rate of organic sedimentation. Conditions favoring organic sedimentation, however, have probably not been at an optimum, and with unconformities representing unknown periods of time in the profiles, this estimate may be low. Also subsidence due to drainage and deflation, augmented by cultivation, has further reduced the thickness of the profiles. As Lower Klamath Lake was born in the Pleistocene, the sedimentary columns may represent most of postglacial time. The occurrence of the artifacts on the fossil lake bed underlying 2 meters of organic sediments, with almost 2 meters of underlying limnic peat, tends to substantiate further this greater age. As the emplacement of the artifacts is probably to be correlated with the warm, dry middle postglacial climate dated from about 8,000 to 4,000 years ago (Cressman, 1940; Antevs, 1946), they lend yet additional support to a substantially greater total age for the sedimentary columns.

The possible age of the volcanic ash stratum in the Washington sedimentary columns is of some significance in the consideration of their chronology. The middle postglacial warm, dry period, as indicated by evidence from several sources, is also revealed by many of the pollen profiles to have been felt in the Pacific Northwest, as will be shown later. The culmination of the climatic maximum apparently occurred some time prior to the time of the volcanic eruption responsible for the volcanic ash stratum. If the xeric period was between 8,000 and 4,000 years ago, as suggested above, the ash in the Washington postglacial sediments was deposited possibly about 6,000 years ago. As mentioned above, the average thickness of the profiles is 7.2 meters, while the average position of the ash is at 4.4 meters. If the total sedimentary column represents about 18,000 years, the average rate of deposition above the ash has been about 1,400 years per meter, and below it has been about 4,300 years per meter, or a ratio of 1 to 3. A range of 1 to 9 meters in the thickness of sediments overlying the ash horizon shows that this ratio is not at all impossible (table 1). If the depositional rate as determined from the total average thickness of the profiles on glacial drift is applied, or 3,600 years per meter of limnic peat and 1,700 years per meter of fibrous peat, a discrepancy appears. As an average of 0.9 meter of limnic and 3.5 meters of fibrous peat overlies the volcanic ash stratum, this would imply a total time of about 9,000 years since the volcanic activity recorded in the sediments, or a difference of 3,000 years between the two methods of computation. It is realized that the rates of organic sedimentation as computed above on the basis of an estimated chronology to begin with can be applied only in a general way. It is felt, however, that the foregoing chronological considerations concerning the sedimentary columns, volcanic eruptions, and the length of postglacial time serve as a practical basis for later chronological consideration of the general trend of forest succession and climate in the Pacific Northwest.

METHODS AND TECHNIQUE
COLLECTION OF PEAT SAMPLES

In the collection of peat samples a Hiller borer was used except in two cases where the Davis sampler was employed. The only advantage seen in the latter is the greater ease in carrying, especially over rugged mountain terrain. It can by no means be used as accurately as the Hiller type. The locating of organic sediments suitable for pollen analysis in the Pacific Northwest is not always easy. The chief obstacles in obtaining sufficient sedimentary columns to provide a representative record of postglacial vegetation over the Pacific Northwest have been the inaccessibility and widely scattered occurrence, as well as the vast area involved. Time and again, after many miles of driving and walking to obtain profiles from sites of hydrarch succession, as indicated by topographic and forestry maps and other sources of information and hearsay, organic sediments were found to be absent or unsuitable for pollen analysis.

Before obtaining sedimentary columns for pollen analysis, a reconnaissance was made of the adjacent terrain to determine the best area to sample in order to obtain a minimum of inorganic material, and the deepest possible profile, and to determine the origin of the basin and the composition of the adjacent vegetation. A series of test borings was made to determine the area of the deepest profile, and the writer obtained much thicker sedimentary columns than Dachnowski-Stokes (1936) on the same sites of hydrarch succession. Borings were made as close to the water's edge as possible where a pond or lake still persisted, in order to have profiles with as nearly as possible the same relative proportions of limnic and fibrous peat from a given area. This procedure also helped to obtain profiles with the volcanic ash in about the same relative stratigraphic position in those bogs containing it. Near the margin the volcanic ash stratum usually occurs in the fibrous peat while nearer the center it is interbedded in the limnic peat.

The interval of sampling the sedimentary column was based on its thickness. In the deeper profiles samples were taken at quarter-meter or 2-decimeter intervals, while in the shallower deposits they were taken at decimeter intervals. In some of the earlier pollen studies in the Pacific Northwest samples were obtained at half-meter intervals (Hansen, 1938, 1939a, 1939b). A half-meter of organic sediments, however, may represent a thousand years or more, and record the culmination of successional and climatic trends or perhaps an entire minor trend. This is more true of shallow profiles known to be equivalent in age to those of much greater thickness. In deeper profiles, located in regions where normal forest succession has occurred, apparently without much influence of climatic changes, such as in the Puget Sound region, pollen analysis of horizons at small intervals reveals no marked difference in the general trends for the area. In some cases it results in a smoother curve, and in others a more saw-toothed effect is produced, but the fluctuations are small and can be attributed to the peculiarities of pollen analysis itself. Theoretically, the limnic peat should be sampled at smaller intervals than the fibrous peat because of its slower rate of accumulation. As previously mentioned, sampling was continued downward into the sand and gravel underlying the organic sediments, and it is believed that in most or all cases a record of the late-glacial and postglacial vegetation is represented by those profiles resting upon glacial drift or its chronological equivalent. In some profiles the analysis of sediments resting directly upon bedrock suggests, with little doubt, the portrayal of the earliest postglacial vegetation in the pollen profiles. Another consideration in the choice of sampling sites was the degree of naturalness of the surface. It is unfortunate that profiles obtained from cultivated bogs and swamps lack the uppermost sediments and, therefore, a record of the present-day vegetation. This is especially true in the Willamette Valley where three out of five profiles lack the latest formed sediments.

The peat samples were placed in 6-dram vials at the time of boring. In most cases a single hole was used in order to avoid depth disparities induced by stratigraphic gradients even within a small area. Samples were not taken from the top of the half-meter core; thus a mixture of sediments was avoided from several higher levels due to their falling or being pushed down by the peat borer. The peat sampler was washed after each core was obtained. If standing water was at hand, the sampler was swished in it; a pail of water and a small whisk broom were found to be the best method if a pool of water was not available. During the process of sampling the type of sediment, unusual characteristics, charred horizons, and thickness of glass or pumice stratum were noted. The sediments were preserved by adding 20 per cent alcohol within a few hours after sampling.

PREPARATION OF SEDIMENTS

In preparation of the peat for microscopic analysis, the potassium hydrate method was used most of the time. This method was modified in some respects at some levels and in some profiles, in order to meet varying conditions of the sediments. In general it consisted of boiling about 2 cc. in a 5 per cent or less solution of KOH for 10 minutes. Gentian violet stain was added before the boiling. Several other stains were tried, but gentian violet was found to be most satisfactory, as it stained practically nothing but the pollen grains. In the lower inorganic sediments of some profiles the pollen content was so sparse that dissolution of the sand, silt, and clay was effected by hydroflouric acid. Sand in peat from dune bogs along the coast was eliminated by decanting; this method was also used in removing the coarser sand from the lower strata of other bogs. After removal of the inorganic material by hydroflouric acid the organic sediments were washed until no trace of an acid reaction remained. Then they were boiled for 5 minutes in 5 per cent KOH in order to present the same conditions as the other method. Marly sediments were treated with nitric acid and then washed until all traces of acid were removed, and the KOH method was applied. After boiling, the coarser types of sediments were strained through cheesecloth or a metal sieve, with several washings. Examination of the residue revealed little or no pollen and the few grains remaining were of the same specific proportions as those used as a basis for the pollen profiles. The boiled and strained sediments were centrifuged and about 0.5 cc. was mixed with warm glycerin jelly. This mixture was permitted to stand in the centrifuge tube in warm water until the pollen grains had absorbed the glycerin jelly, and then two drops were mounted on a slide under a $\frac{7}{8}$ inch cover glass. The rest of the mixture was preserved in a stoppered vial for future reference if necessary. After centrifuging, the liquid was repeatedly examined for pollen, but none was ever found. The tap water used in boiling and washing was also examined for pollen, as its source was usually an open reservoir in the forested mountains. In only a few instances was pollen noted. Containers were kept covered in order to prohibit the entrance of atmospheric pollen, a precaution found to be very important during the anthesis period for conifers. An effort was made to maintain a fairly constant procedure, quantitatively and qualitatively, in the preparation of the pollen-bearing sediments, so as to have a constant basis for comparison. Obviously, in those cases where some of the sediments were removed, either by dissolution or straining, the pollen frequency on the slides is not correlative with the original quantity of the sediments. Various other methods in preparation of organic sediments for microscopic analysis were used (Geisler, 1935; Erdtmann, 1936; Assarsson and Granlund, 1924), but no advantage could be seen in these over the potassium hydrate method. Pollen grains are generally abundant in Pacific Northwest organic sediments, and only at a few levels were they too few to use in the pollen profiles. They are much more abundant than in organic sedimentary columns from Wisconsin, which the writer has analyzed (Hansen, 1937, 1939c).

NUMBER OF POLLEN GRAINS COUNTED

In earlier fossil pollen studies 200 or more pollen grains of indicator species were identified from each level. In later work 150 grains from each horizon were deemed sufficient to give a representative picture of adjacent vegetation. In some levels where pollen was scarce 100 grains were counted. In a few levels only 50 grains were found after considerable microscopic work, but in no case have the pollen proportions at a given level been computed upon the basis of less than 50 pollen grains of indicator species. In the lower sandy sediments pollen is so scarce that some of these horizons have not been included in the pollen diagrams.

It has been noted consistently in pollen studies in the Pacific Northwest that the proportions of the indicator species change very little as the number of pollen grains identified goes beyond one hundred. The number counted varies with different workers, but it is generally conceded that not less than 100 grains of forest tree species or other indicators should be counted for each level. Bowman (1931) believes that 1,000 grains are necessary per sample, but few or no workers have counted this many. Barkley (1934) concluded by statistical methods that a count of 175–200 grains should be adequate for reliable representation.

Pollen of non-indicator species is common at certain levels, and as many as 600 grains have been counted. Red alder is by far the most abundantly represented of the non-forest tree species. Fern spores are also numerous, and those of bracken fern (*Pteridium aquilinum*) are the most common. In addition to pollen and spores, other microfossils that have been noted most commonly are diatoms, desmids, protozoan tests, sponge spicules, and insect eggs and parts.

IDENTIFICATION OF FOSSIL POLLEN

The identification of the pollen preserved in the sedimentary column is, perhaps, one of the most critical phases in the method of pollen analysis. Most workers, both in Europe and in North America, have concluded that when possibly more than one species of a given genus are represented by their pollen, accurate differentiation cannot be attained.

In certain families it is not feasible to attempt separation of even the genera by their pollen. The segregation by their fossil pollen of the species of certain genera is important, however, because one species may express entirely different climatic trends from another, or provide evidence for an unsuspected paleic distribution. When a pollen profile for a genus is considered, a true picture of the vegetation history and its indicated climatic trends may not be presented. It becomes extremely important, therefore, that when it is certain more than one species of a genus is represented by its pollen, and that they reflect different paleoecologic conditions, every attempt should be made to segregate their pollen. This is shown by pollen analysis of buried soils in the Piedmont near Spartanburg, South Carolina (Cain, 1940, 1944), in which there is evidence of a prehistoric distribution of *Abies fraseri* and *Pinus banksiana* different from that of the present.

Although it is not difficult to separate the three common genera of conifers with winged pollen grains, *Abies*, *Picea*, and *Pinus*, it is hard to segregate the species of each genus, and impossible for some species. Erdtman (1931) says it is possible to distinguish the pollen of *Picea canadensis* from that of *P. mariana*, and that of *Pinus banksiana* from that of *P. murrayana*. Wilson (1938) and Wilson and Kosanke (1940) separate *Picea glauca* from *P. mariana*, and *Pinus banksiana* from *P. strobus* and *P. resinosa* in both post-Wisconsin and pre-Kansan profiles. Hormann (1929), Stark (1927), and Bertsch (1931) determined pollen size-frequency distributions of species of pine and birch, and applied them to fossil pollen studies, while Jaeschke (1935) and von Sarntheim (1936) published criticisms of some of this work and concluded that the method was unreliable. Deevey (1939) determined pollen size-frequency distributions for *Pinus banksiana*, *P. strobus*, *P. resinosa*, and *P. rigida*, and decided that their application to the identification of fossil pollen was infeasible. Cain (1940) determined the size-frequency distributions for twelve species of pine distributed in eastern and southern United States. Size-frequency distribution of fossil pollen revealed a trimodal curve similar to that of three species of modern pollen, and it was assumed that the species represented were the same as those that furnished the basis for the modern pollen size-frequency curves.

IDENTIFICATION OF FOSSIL POLLEN IN THE PACIFIC NORTHWEST

In preparation for pollen analysis of sedimentary columns in the Pacific Northwest it was immediately realized upon examination of modern pollen that the species of certain groups were going to be difficult or impossible to segregate. It was found that the species of the genera *Abies* and *Pinus* offered the greatest difficulty because of the several species involved, their morphological similarity, and somewhat similar size-frequency distribution. Fresh pollen was obtained of all species that are potentially represented in the peat profiles of the Puget Sound region. One of the first things noted in its study was a size differential of a given species when different methods of preparation were used. An especially significant increase in size was noted when the pollen was treated with KOH, which resulted in some confusion in identification of fossil pollen. The Sphagnum peat of the Puget Sound region is unusually raw and well preserved, and treatment with potassium hydrate did not seem to be necessary to deflocculate the peat for microscopic analysis. In view of these results all fresh pollen was treated with KOH and mounted in glycerin jelly, and the KOH method has been consistently used in the preparation of the sediments in order to avoid size differential between modern and fossil pollen of the same species. Cain (1944) has since published evidence that treatment of fresh pollen grains of *Abies fraseri* with the acetolysis and KOH methods resulted in a grain swelling of 8.5 per cent over that when water or alcohol was used.

Before pollen analyses were made of sedimentary columns in the Puget Sound region (Hansen, 1938), size-frequency distributions and mean dimensions were determined of the pollen grains of lodgepole pine (*Pinus contorta* Dougl.), western white pine (*P. monticola* Dougl.), whitebark pine (*P. albicaulis* Engelm.), western yellow pine (*P. ponderosa* Dougl.), lowland white fir (*Abies grandis* Lind.), noble fir (*A. nobilis* Lind.), silver fir (*A. amabilis* (Dougl.) (Forbs), alpine fir (*A. lasiocarpa* (Hook) Nutt.), Sitka spruce (*Picea sitchensis* (Bong.) Carr.), and Engelmann spruce (*P. engelmanni* (Parry) Engelm.). The fresh pollen was treated with potassium hydrate and mounted in glycerin jelly. One-hundred pollen grains of each species were measured as to length of grain and overall length. In the separation of fossil pollen the grain was measured with a lens combination of 430×, and then listed as that species within whose size-range it fell. If the dimension was in the zone of overlap of two species, it was listed as an unknown. Distorted pollen or that lying in an unmeasurable position was also included in the unknown. No size-frequency distribution curves of the modern pollen were compared with those of the fossil pollen. As the area of investigation was expanded in the Pacific Northwest into different phytogeographic areas, fresh pollen of most of the forest tree species was obtained, size-frequency distributions determined, and the size-range method applied to analyses of the sedimentary columns. Additional species of winged conifer pollen included sugar pine (*Pinus lambertiana* Dougl.), knobcone pine (*P. attenuata* Lemm.), and white fir (*Abies concolor* (Gord.) Engelm.).

Before pollen analyses of about twenty-five additional profiles included in this study were made,

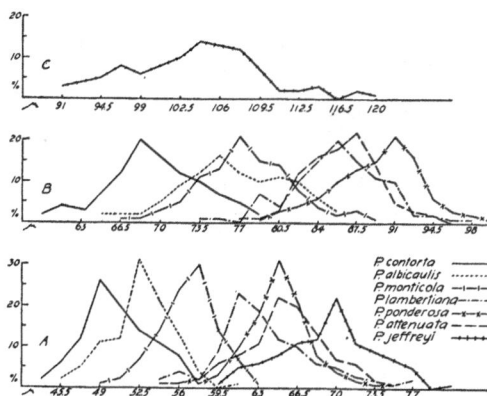

FIG. 17. Size-frequency distribution curves of modern pine pollen. *A.* Size-frequency of cell only. *B.* Size-frequency of over-all length, including air sacs. *C.* Size-frequency distribution of over-all length of *Pinus jeffreyi* pollen.

however, new slides of all of the species of winged conifer pollen were made. This pollen was treated with potassium hydrate and mounted in glycerin jelly and size-frequency distributions determined. The pollen of these species represented new collections from different localities for the purpose of determining whether or not the pollen size of a given species varied in different areas. The results of these new measurements are shown by size-frequency distribution curves (fig. 17). With respect to pine, the new size-frequency distributions were very similar to the first and offered no basis for reconsideration of the analyses of some forty profiles already made. In the case of fir, however, certain discrepancies were noted between the first and second studies which are significant and will be discussed later.

PINUS

There is an appreciable size-range difference, but with some overlap, between lodgepole, western white, and western yellow pine pollen. The size-range of whitebark pine, however, is within those of lodgepole and white pine, largely the latter (fig. 17*A*). The size-frequency distributions of sugar and knobcone pine are largely within those of yellow pine. The separation of yellow pine fossil pollen, however, is fairly easy, owing to the larger bladders in proportion to the size of the cell than in the other two species (fig. 17*B*). Furthermore, the range of sugar and knobcone pine in the Pacific Northwest is restricted to southern Oregon and they are not the preponderant species in the forest complex. Confusion with these species, then, is possible only in profiles located in the southern Cascade Range of Oregon. The range of whitebark pine near timberline also tends to eliminate some of the confusion in separating its pollen from that of lodgepole and western white pine, while yellow pine can readily be separated from these three species by its larger size, as well as by its comparatively larger bladders. At those levels in the sedimentary columns where pine is represented by over 50 per cent of the forest tree pollen, 100 pollen grains of pine have been identified, and those listed as unknown have not been considered in the construction of the pollen diagrams or its interpretation. At levels where pine pollen represents less than 50 per cent of the total forest tree pollen, then only the pine pollen included in a total of 150 forest tree pollen grains was identified. While the presence of unknown pine pollen, sometimes as great as 20 per cent, must be considered as a source of error, it seems probable that the species represented by the unknown pollen would be in the same proportions as that listed as known. Thus, if 75 per cent of the pollen is identified as that of lodgepole pine, then 75 per cent of that listed as unknown should be that of lodgepole pine, if the size-range method of identification is reliable.

The present range of the several species of pine in the Pacific Northwest is of value in supporting or refuting the results of fossil pollen identification by the size-range method. Possible prehistoric distribution must also be considered in evaluating the results of pollen identification by this method. In Washington and Idaho, the only species of pine that have probably existed within range of pollen dispersal to the sites of the sediments during postglacial time are lodgepole, western white, western yellow, and whitebark pine. In the Puget Lowland lodgepole and white pine are the chief species of pine at present, with the latter by far preponderant. It is, therefore, presumed that they have been most abundant during postglacial time because there is no evidence that yellow or the other species of pine having larger pollen grains were ever present. Only a few pollen grains noted in the sediments have been assigned to yellow pine. Upon the basis of their slightly different size-frequency distributions, lodgepole and white pine can probably be separated with a fair degree of accuracy.

In eastern Washington, lodgepole, white, and yellow pine are the most abundant species and the identification of their pollen by the size-range method has provided pollen profiles of each that indicate logical postglacial successional trends. In profiles from higher altitudes the pollen of whitebark pine is undoubtedly present and listed with that of both lodgepole and white pine. In some cases it probably drifted down from higher elevations, while in others it was derived from whitebark pine that existed at lower elevations during the cooler climate of early postglacial time. As white and whitebark pine respond somewhat similarly to the same climatic trends, their pollen considered collectively probably does not result in misinterpretation of the pollen record as far as climate is concerned.

In the coastal strip lodgepole pine is the only pine species noted in the forest complex. As the lodgepole pine forests are usually located windward to the bogs, there is little chance for other species of pine farther inland to be represented. Although lodgepole pine on the coast may differ taxonomically from that farther inland, the size-frequency distribution of their pollen does not seem to differ any more than that of pollen from a single species. In the Willamette Valley the only species of pine known to occur naturally is yellow pine in the southern part. This species is sparsely represented in some of the sedimentary columns, but insufficiently to warrant interpretation of climatic or other trends. The most abundantly represented species have been identified as lodgepole and white pine, and their pollen profiles indicate feasible trends of succession.

Sedimentary columns located in the Cascade Range of southern Oregon potentially have the greatest number of pine species represented by their pollen in the Pacific Northwest. Here it is possible to have all species listed above represented. In addition the present ranges of Jeffrey pine (*Pinus jefferyi* Grev. and Balf.) and foxtail pine (*P. balfouriana* Murr.) (Munns, 1938) are such as to be possibly represented. However, the most abundant and widely distributed species are white, lodgepole, whitebark, yellow, and sugar pine, and these are probably all represented in the sedimentary columns. The most common of these are white, lodgepole, and yellow pine. The pollen of the first two can be separated, but undoubtedly includes more or less of that of whitebark pine, depending upon the elevation of the site of the sediments. The pollen of yellow pine can probably be separated from that of sugar and knobcone pine, but this has not been done because the value would not be commensurate with the work involved.

The author does not pretend that segregation of pine fossil pollen by the size-range method is infallible or statistically accurate. In view of the indicator value of some of the species, however, it seems better to separate them by this method than to consider them all collectively. The latter would provide little basis for the interpretation of the forest succession and climate in the vicinity of each sedimentary column. Neither would it furnish any means of correlation with other profiles within the same phytogeographic province nor with those of several phytogeographic provinces. The consistency and logic of the indicated trends of the pollen profiles in the several areas, even though portrayed by different species, supports the feasibility of the size-range method in separating pine fossil pollen.

A comparison of the size-frequency distribution of fresh pollen and that of fossil pollen has been shown by Cain (1940) to be of some value in determining the probable identity of the fossil pine pollen. The author has merely measured the pollen grain and assigned it to that species within whose size range it fell, without keeping a record of the measurements. In certain subsequent pollen analyses a record of the dimensions was kept for the fossil pine pollen at certain levels and a size-frequency distribution curve constructed. This was done for levels where a sharp change in the relative proportions of the several species of pine occurred, in profiles where it was reasonably certain that only a single species of pine was involved, or for several levels where there were gradual changes in the proportions of pine pollen upward in the profile. In some cases, these curves are shown for levels of profiles for which the results have already been published.

In a 7.3-meter profile located near Spokane, Washington (Hansen, 1939b), remeasurement of the fossil pine pollen at three selected levels shows interesting size-frequency distribution curves closely correlated with the curves for modern pollen of lodgepole, white, and yellow pine. The pollen profile of lodgepole pine reveals a gradual decline from predominance at the lowest level to small proportions at the top. White pine declines to the middle of the profile and then slightly increases to the top, while yellow pine is absent at the bottom and expands to predominance about half-way up and maintains this status to the top. One hundred pine pollen grains were measured from each, the bottom level, the 3.5-meter level, and the uppermost level. Comparison of the size-frequency distribution with those of modern pollen reveals close correlation with respect to the three modes (fig. 18), which suggests that the size-range method

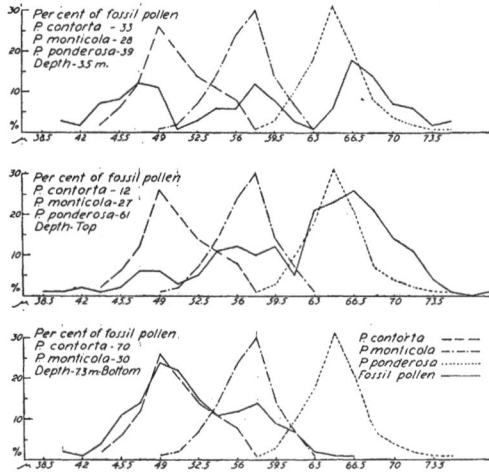

FIG. 18. A comparison of the size-frequency distribution of modern pine pollen with that of fossil pine pollen at three levels of a profile from near Spokane, Washington. In this and the following similar diagrams the percentage of fossil pollen is based on the size-frequency distribution of 100 grains at each level.

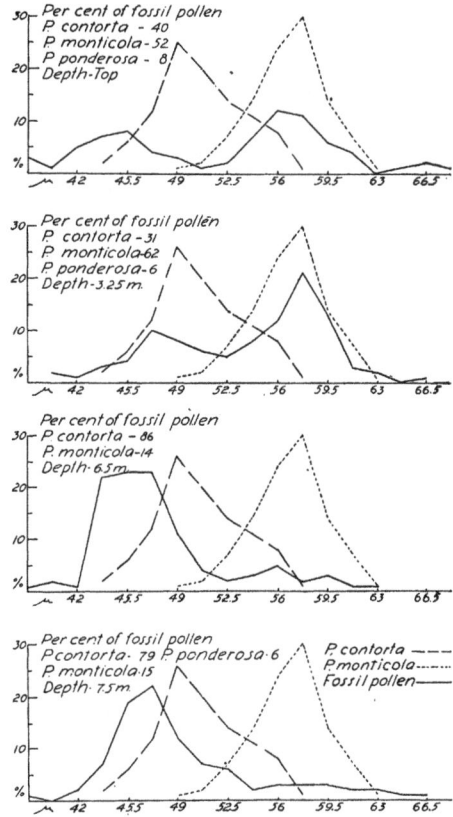

FIG. 19. A comparison of the size-frequency distribution of modern pine pollen with that of fossil pine pollen at four levels of a profile from near Bonners Ferry in northern Idaho.

for fossil pine pollen identification is reliable in depicting the major trends.

In order to test further the size-range method by comparisons of size-frequency distribution of modern and fossil pollen, the same procedure has been applied to several sedimentary columns in which the trends of pine succession are distinctly different. The comparative successional trends between lodgepole and white pine in a profile near Bonners Ferry in northern Idaho, as determined by the size-range method, are substantiated by the comparisons of the size-frequency distribution curves of modern and the fossil pollen (Hansen, 1943e), In a 7.5-meter sedimentary column, lodgepole pine is recorded to predominance for more than half of postglacial time and is then superseded by white pine, as interpreted from the pollen profiles based on the size-range method. The measurement of 100 pine pollen grains at each of four levels and the results embodied in size-frequency distribution curves reveal variation in modes reflecting the relative change in the pollen proportions of the two species as postglacial time progressed (fig. 19).

Size-frequency distribution curves for two levels, one just below and the other immediately above a pumice stratum in a 6.5-meter profile near Bend, Oregon, show the radical change in forest composition due to the influence of the pumice mantle in adjacent regions (fig. 20). A sharp contrast in proportions of lodgepole and white pine is disclosed by size-frequency

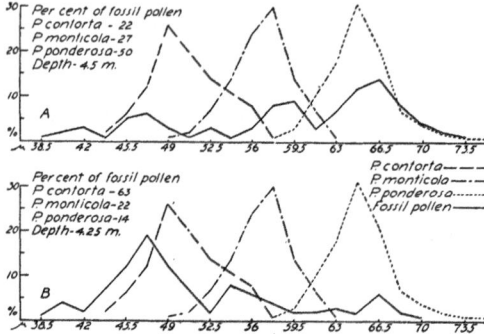

FIG. 20. A comparison of the size-frequency distribution of modern pine pollen with that of fossil pine pollen at two levels of a profile from near Bend (Tumalo Lake), Oregon. A stratum of Mount Mazama pumice occurs just below the 4.25 meter horizon. *A.* Size-frequency distribution before the pumice fall. *B.* Size-frequency distribution after the pumice fall. The rapid expansion of lodgepole pine suggests that the pumice fall was detrimental to the persistence of western yellow pine.

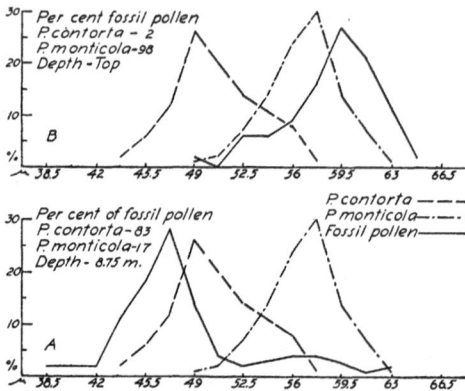

FIG. 21. A comparison of the size-frequency distribution of modern pine pollen with that of fossil pine pollen at two levels of profile from near Poulsbo, Washington. *A.* The bottom level reveals the early postglacial predominance of lodgepole pine. *B.* The surface level shows the distortion effected by the presence of two western white pine trees growing on an island near the center of the bog. This bog is located in a Douglas fir forest.

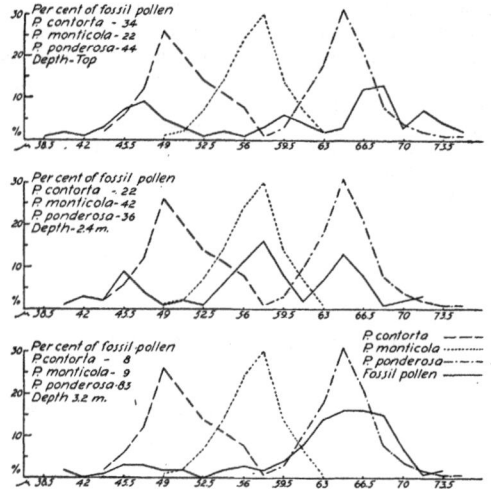

FIG. 22. A comparison of the size-frequency distribution of modern pine pollen with that of fossil pine pollen at three levels of a profile located in the upper Rogue River Valley, west of Crater Lake, Oregon. The influence of the possible representation of at least five species of pine is reflected in the sprawling curves with more than three modes.

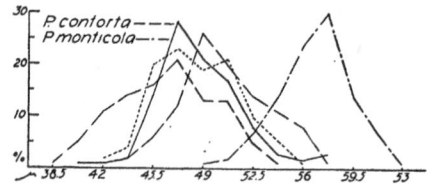

FIG. 23. A comparison of the size-frequency distribution of modern pine pollen with that of fossil pine pollen at three levels of a profile from near Newport, Oregon, on the Pacific Coast. The only pine known to occur in the vicinity is lodgepole, and a forest of lodgepole grows on the mature bog surface.

distribution curves in an 8.75-meter profile near Poulsbo, Washington, for the bottom and top levels. In the lowest level lodgepole pine is represented by 83 per cent of the pine pollen, while at the surface white pine is preponderant with 98 per cent of the pollen (fig. 21). The high proportions of the latter at the top are due to the presence of two large white pine trees on a knoll near the center of the bog. Size-frequency distribution curves for three selected levels from a 4.4 meter sedimentary column from the upper Rogue River Valley in southern Oregon disclose three principal modes and certain irregularities reflecting the probability of lodgepole, white, and yellow as being the chief contributors and whitebark and sugar pine as minor contributors (fig. 22). Finally, the size-frequency distribution curves for pine from several levels of a bog near Newport, Oregon, on the coast, disclose a consistent similarity with that of modern lodgepole pine pollen (fig. 23). The bog is in the climax stage, and a forest of lodgepole pine grows on the surface.

ABIES

In the more recent pollen size-frequency determinations of the several species of fir in the Pacific Northwest some discrepancies have been noted between the results of this and the earlier study, which at present makes the identification of fir fossil pollen infeasible. The study will be carried further, however, using more and different collections of fresh pollen in order to obtain sufficient statistical data to indicate with greater certainty to what degree the size-range method may be used for identification of fir fossil pollen. As yet no modern pollen of red fir (*Abies magnifica* Murr.) has been observed.

The results of both studies revealed very similar size-frequency distributions for alpine fir which has the smallest pollen of the Pacific Northwest firs. Of five Pacific Northwest species studied by Wodehouse (1935), alpine fir was also found to have the smallest pollen grain. In the first study the other species observed in order of their size, beginning with the smallest, were lowland white fir, silver fir, noble fir, and white fir. The size-frequency distribution of silver fir lay between that of lowland white and noble fir, and could not be separated in postglacial sediments. Wodehouse found that lowland white fir has a pollen grain similar in size to that of alpine fir, while the other three that he observed, in order of their size from the smallest, are white fir, red fir, and noble fir. The second study of size-frequency distribution on the basis of 100 grains, disclosed that the mean of silver fir was larger than that of either lowland white or noble fir, and that the size-frequency distributions of all three are somewhat similar and cannot be used to segregate these three species in postglacial sediments (fig. 24). The second study of white fir pollen again

revealed that it is the largest fir pollen of the Pacific Northwest, but a greater size-range was noted than in the first study (fig. 24). Its size range lies outside that of any of the other species. Thus it seems that at present only alpine and white fir can be separated from the others with any degree of accuracy. The distinction of alpine fir is made simpler by the bladders which are larger in proportion to the size of the cell than in any of the other firs.

The proportions of true fir pollen preserved in postglacial sedimentary columns in the Pacific Northwest are usually too low to warrant interpretation of climatic trends, although in some profiles they offer support for the trends interpreted from the fluctuations of other species. It is probable that in the lowland profiles lowland white fir is the most abundantly represented by its pollen. In montane profiles on the west slope of the Cascade Range, noble and silver firs are perhaps more strongly recorded. In the southern Cascades of Oregon, all of the Pacific Northwest species of true fir may be represented in the sedimentary columns.

The greatest and most consistent representation of fir is in the Rogue River profile located southwest of Crater Lake. The principal species in the adjacent forests are white and alpine fir, and measurement of the pollen at several levels of the sedimentary column reveals a considerable size-range, with the strongest modes of the size-frequency distribution curves centered near the means of alpine and white firs. No further attempt has been made, however, to separate the species of fir fossil pollen in the twenty-five profiles of this study for which the pollen analytical data have not been published. The use of the size-range method to segregate fir fossil pollen interred in Pacific Northwest postglacial sedimentary columns will have to be held in abeyance until further studies of modern pollen are carried out. On the basis of the size-frequency distributions determined up to the present, however, it seems improbable that this method will be feasible in separating all species of fir pollen.

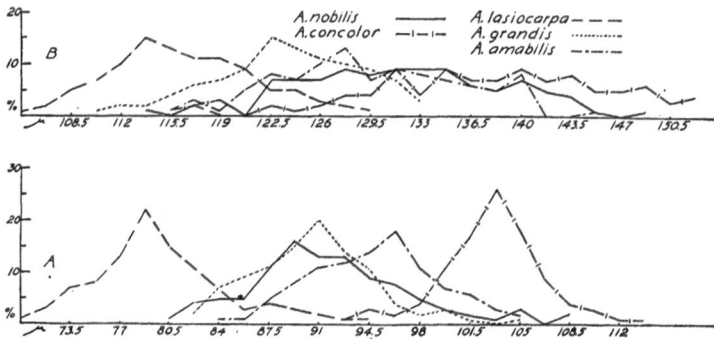

FIG. 24. Size-frequency distribution curves of modern fir (Abies) pollen. *A*. Size-frequency of cell only. *B*. Size frequency distribution of over-all length, including air sacs.

PICEA

The three species of spruce that grow in the Pacific Northwest are Sitka spruce, Engelmann spruce, and weeping spruce (*P. breweriana* Wats.). The first is confined largely to the coastal strip and it is probably the only species of spruce represented in the sedimentary columns from this region. Engelmann spruce is a montane species and is represented in inland bogs. In the Puget Sound region it is possible that both species are represented in the organic sediments. Weeping spruce is localized in southwestern Oregon and its pollen has not been observed. The size-frequency distributions of Sitka and Engelmann spruce pollen do not overlap to any great extent and can be used as a means of segregating their fossil pollen (fig. 25). Spruce is not represented in significant proportions except in sedimentary columns from along the coast, and size-frequency distributions of the spruce fossil pollen from a few levels indicate that Sitka spruce is the species represented (fig. 26). In other profiles, even though located within forest areas containing an abundance of Engelmann spruce, this species is sparsely represented. Observation of the species in the field suggests that it is not a prolific pollen producer and it may be under-represented in the sedimentary columns.

OTHER FOREST TREE SPECIES

Several other genera of forest trees in the Pacific Northwest are represented by a single species, and the identification of their fossil pollen offers no difficulties. In a few other genera the segregation of specific pollen is simplified by morphological differences or because of the relative geographic distribution of the species concerned.

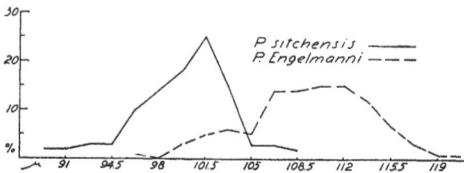

FIG. 25. Size-frequency distribution curves for modern spruce pollen.

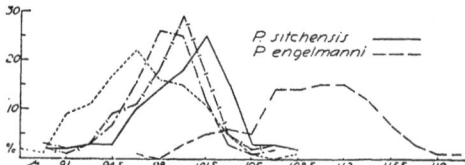

FIG. 26. A comparison of the size-frequency distribution of modern spruce pollen with that of fossil spruce pollen at three levels of a bog at Woahink Lake, located near the Oregon coast. Sitka spruce is the only known spruce in the vicinity.

Douglas fir (*Pseudotsuga taxifolia* (Poir) Britt.), one of the most important forest tree species in the Pacific Northwest, is readily identified by its fossil pollen with no possible source of error existent. It is the only species of the genus in the Pacific Northwest, and it has a large spherical pollen grain with no counterpart in the sedimentary columns.

Western hemlock (*Tsuga heterophylla* (Rafn.) Sarg.) is another common forest tree that is readily segregated by its fossil pollen. The shape of the pollen grain varies from lens-shaped to spheroidal. The surface of the exine is heavily reticulate-corrugated, and some grains show a slight swelling suggesting rudimentary bladders.

The other species of the genus, mountain hemlock (*T. mertensiana* (Bong.) Carr.) has an entirely different type of grain. It is smaller and has well developed bladders similar to those of *Pinus*, *Abies*, and *Picea*. The pollen of mountain hemlock is readily distinguished from that of these other genera by the much finer reticulum on the bladders, and the reddish stain effected by gentian violet.

Western larch (*Larix occidentalis* Nutt.) has a grain somewhat similar to that of Douglas fir. It is spheroidal, has a thinner exine than that of Douglas fir, and is smaller in size. The accurate separation of its pollen in Pacific Northwest sedimentary columns seems to offer no problem and involves little source of error. The chief complication is its possible confusion with that of alpine larch (*L. lyallii* Parl.). Whereas the pollen of these two species is similar, the restricted range of the latter, near timberline and in scanty proportions, suggests that it has never played an important role in postglacial forest succession.

Oregon white oak (*Quercus garryana* Dougl.) is an important species in the Willamette Valley, and is significantly represented by its pollen in this area and to a lesser degree in a few profiles from the "Tacoma Prairies" south of Puget Sound. It is the only species of the genus in the Pacific Northwest, and there is little difficulty in distinguishing its pollen in the sedimentary columns. Its pollen grain closely resembles that of red oak (*Q. borealis* Michx.) of north central and eastern United States, and stains a scarlet color with gentian violet.

There is a group of conifers whose pollen is poorly preserved in Pacific Northwest sedimentary columns and whose pollen is also somewhat similar in appearance. These conifers include western red cedar (*Thuja plicata* Don), Port Orford cedar (*Chamaecyparis lawsoniana* (Murr.) Parl.), Alaska cedar (*C. nootkatensis* (Lamb.) Spach.), incense cedar (*Libocedrus decurrens* Torr.), redwood (*Sequoia sempervirens* (Lamb.) Endl.) and western juniper (*Juniperus occidentalis* Nutt.). The present distribution of these species denotes that they would all probably be represented by their pollen in one or more of the sedimentary columns if their pollen were preserved. Fresh pollen of all species except redwood has been secured. Western red cedar has the widest geographic range of these species in the Pacific Northwest (Munns, 1938), and if its pollen were preserved it would undoubtedly be the strongest represented. Pollen identified as that of this species has been noted in small proportions in certain profiles. In sedimentary columns from the semi-arid regions the absence of pollen of these species is unfortunate, as the postglacial trends of some would probably support trends depicted by the represented species.

MISCELLANEOUS SPECIES

In addition to the forest trees whose pollen records reveal the course of postglacial forest succession and climatic trends, pollen of other species has been noted and recorded during microscopic analysis. Among these, the most abundantly represented is red alder (*Alnus rubra* Bong.), while bigleaf maple (*Acer macrophyllum* Pursh.), and Oregon ash (*Fraxinus oregona* Nutt.) are recorded to some extent, depending upon the location of the sediments. Salicoid pollen, probably from various species of willow and poplar, may be abundant at certain levels, but no attempt has been made to segregate these as to species. To the east of the Cascade Range alder pollen present is probably from white alder (*Alnus rhombifolia* Nutt.) or mountain alder (*A. tenuifolia* Nutt.), depending upon the location of the sediments.

Fresh pollen has been collected from many species of hydrophytic plants and their fossil pollen record reveals the progress of hydrarch succession. Fresh pollen of many species of grasses, Composites, and Chenopods has also been collected and studied. These collections are small in comparison to the large number of possible species represented in the postglacial sediments, particularly those from the treeless areas of eastern Oregon and Washington. The pollen grains of these families are distinct from one another, but no attempt has been made to segregate the species. In view of the large number of species involved, especially the grasses, separation of specific fossil pollen would be impossible.

SUMMARY

One of the most critical phases of fossil pollen analysis is the identification of the pollen grains preserved in the sediments. In the Pacific Northwest the separation of the several species of pine by their pollen offers the greatest problem. Pollen of most of the other conifers represented in the sedimentary columns can be accurately segregated with the exception of true fir. Pending further investigation to determine a reliable method of separating fir species pollen, it must be considered collectively as a genus. This treatment is not critical in this study because fir pollen is not abundant in most profiles. No attempt has been made to separate the specific pollen of

grasses, Chenopods, and Composites, but their consideration as families seems to be satisfactory as far as interpretation of climatic trends is concerned.

In separating the species pollen of pine where it was evident that more than one species was represented, the size-range method was employed. Although the size-ranges of some species overlap, and those of others are similar, the most important species are separable on this basis. Lodgepole, western white, and western yellow pine, the most important indicator species, overlap to some degree, but not sufficiently to impair the use of the size-range method. Other species of pine that are largely inseparable are of limited geographic range and not abundant in the forest complex, and thus do not significantly influence the accuracy of the overall results. The comparison of the size-frequency distribution of fossil pine pollen at certain levels in some of the sedimentary columns with those of fresh pollen of the species apparently represented reveals a remarkable correlation and offers considerable support to the reliability of the size-range method in the identification of fossil pollen.

CLIMATE OF THE PACIFIC NORTHWEST [3]

GENERAL STATEMENT

The states of Oregon and Washington range from 42° to 49° north latitude, and comprise an area of over 162,000 square miles. Perhaps no region of similar size in the United States has so wide a variety of local climates. The marine influence on the west and the continental influence on the east, together with the mountain barriers which not only prevent free interchange of air but greatly affect the temperature and moisture of such air masses as do surmount them, result in striking contrasts. That part of the area lying west of the Cascades has a milder climate than that of any other section of the continent in the same latitude. Some localities on the west slope of the Coast Range and Olympic Mountains have the heaviest annual precipitation in the country, while to the east of the Cascade Range some areas have less rainfall than any part of the United States east of the continental divide. The proximity of the Pacific Ocean and the prevailing westerly winds provide basically a marine type of climate for the Pacific Northwest. The principal features are a small annual range in temperature for the latitude, considerable fog, prolonged cloud-

---

[3] All data used in computing the average precipitation and temperatures have been obtained from the Climatic Summary of the United States for Washington, Oregon, and northern Idaho, U. S. Department of Agriculture, Weather Bureau, up to and including 1930. The averages were brought up to date in most cases with data from the annual Climatological Data, U. S. Department of Agriculture, Weather Bureau, up to and including 1940. Data accumulated since 1940 were obtained from the same publication, U. S. Department of Commerce, Weather Bureau.

iness, an abundant precipitation occurring during the rather mild winter, a relatively dry and cool summer, a long frost-free season, and wind off the ocean during most of the year. The best expression of a marine climate in the area is along the immediate coast, and it then shows increasing modification inland in direct proportion to increasing distance from the ocean. In addition to distance from the coast, differences in altitude, mountain barriers, and local topography further modify the marine climate. These modifications in general tend toward aridity and continentality except on the windward slopes of the mountains, so that many of the elements of a marine climate are lost near the eastern border of the area. Yet eastern Oregon and Washington have a much milder climate than the northern Great Plains. The Cascade Range acts as a partial barrier for the western part of the area against the continental influences of eastern Oregon and Washington, both the heat of summer and the cold of winter, while the Rocky Mountains form yet a second defense against the stronger continental climate of northern interior plains.

There is a large number of climatic sub-types in the Pacific Northwest, some of them highly localized. After the moisture-laden winds from the Pacific Ocean reach the land, the chief controls are the mountain barriers that lie in a general north-south direction athwart the trajectory of air flow. Precipitation is heaviest near the ocean and on the windward slope of the mountains, and the lightest in the deeper valleys and on leeward slopes. Condensation of the moisture is accelerated by dynamic cooling when the air rises to cross the mountains and is diminished by warming through compression as it descends the leeward slopes. The wide variation in climate is well illustrated by a range in the average annual precipitation from 6 inches in south central Washington to 146 inches in the foothills on the southwest slopes of the Olympic Mountains, a range greater than that found in any other state. The wide range in rainfall within a short distance is further shown by an average annual precipitation of 17 inches at Sequim, Washington, at sea level in the northeast corner of the Olympic Peninsula in the rain shadow of the Olympic Mountains, and of 125 inches at Spruce on the west slope of the Olympics, only 60 miles distant. Most of the precipitation in the western part of the area occurs during the winter months, from 75 to 85 per cent of it during the six months beginning October 1 and ending March 31. East of the Cascades greater continentality prevails in this respect, and the winter precipitation ranges from 55 to 75 per cent of the annual total. Nearly all of the precipitation in the Pacific Northwest is caused by the movement of low pressure areas from the north Pacific Ocean eastward over the continent. As summer approaches, however, the storm tracks move northward beyond the area, so that little cyclonic rain falls, while in the winter the southward

movement of the paths of the storm centers causes them to pass directly over. Moreover, the storms are more frequent in the winters. For these reasons heavy precipitation results in the winter season.

The climate west of the Cascade Range is much more equable than that to the east, with much smaller ranges of both diurnal and seasonal temperatures. Along the coast the range between the average annual maximum and average annual minimum temperature is about 13°, while to the east of the Cascades it is over 25°. East of the Cascades the precipitation is light, relative humidity is low, evaporation is rapid, and sunshine is abundant, while to the west conditions are reversed. Snowfall is light west of the Cascades, increasing with elevation on the west slope, at least to moderate levels, and sharply decreasing on the leeward eastern slope and out into the middle Columbia Basin. At Paradise Inn on Mount Rainier, at an elevation of 5,505 feet, the average annual snowfall is almost 50 feet, and in the Olympic Mountains it is probably considerably greater. The prevailing winds throughout the year are westerly, generally blowing from the southwest during the winter and from the northwest in the summer. Thus the marine influence is greater upon the interior east of the Cascades than the continental influence of the middle Columbia Basin is upon the area to the west of the Cascades. When east winds prevail the weather becomes dry in the western part, and in winter the temperature drops below average, while in summer it rises above average. These periods, however, last only a few days. The Columbia Gorge, cut through the Cascades nearly to sea level, serves as a channel for such flow of continental air to the area west of the Cascades.

The interpretation of postglacial climate from postglacial forest succession must be based upon the present climate and distribution and type of vegetation. In the Pacific Northwest the great variation in topography and the wide range of climate with much localization have produced a number of distinct vegetation types. As the postglacial vegetation succession has likewise differed in the several physiographic and climatic provinces, it is necessary to discuss briefly the pertinent climatic characteristics of each province in order to interpret the past climate upon the basis of the present.

COASTAL STRIP

The zone between the ocean shore and the Coast Range of Oregon and Washington, here called the coastal strip (map 1), has the most equable climate in the Pacific Northwest, with low extremes of temperature and precipitation. This strip is only a few miles wide in most places but extends north and south for a distance of almost 500 miles. Its climate is an excellent example of a marine type, as the precipitation and temperature are little affected by altitude or other modifying factors. Increase in altitude in the Coast Range immediately to the east is reflected by a much greater precipitation. The mean annual rainfall at stations on the coast ranges from 94 inches at Tillamook, Oregon, to 60 inches at North Head near the mouth of the Columbia River in Washington. There is a general decrease in rainfall southward along the Washington coast, then a slight increase south of the Columbia River, followed southward again by a decrease, to a degree which remains almost constant, to the southern boundary of Oregon The southernmost station on the ocean, at Brookings, has a mean annual rainfall record of 75 inches. The average of the mean annual precipitation for 9 stations immediately adjacent to the ocean is 76 inches. Only about 15 per cent of the rainfall occurs during the growing season, but this is probably not a limiting factor in plant growth because of the large total rainfall and the high relative humidity prevailing throughout the year. The average annual temperature is about 50°, varying by only a few degrees from north to south. The average daily maximum temperature for 7 stations is about 57°, while the average daily minimum is 44°. Heavy fogs are frequent, snowfall is rare, and the frost-free season is about 250 days long, or even as much as 300 days at some stations. The climate of the coastal strip is classified as "wet, microthermal, with adequate precipitation at all seasons" (Thornthwaite, 1931).[4]

COAST RANGE AND OLYMPIC MOUNTAINS

A few miles inland from the ocean the influence of cooling by the land and of increasing altitude results in the heaviest annual precipitation in North America. The annual rainfall ranges from 53 to 120 inches, depending upon the altitude and upon the location of the station with reference to windward or leeward slopes. The greater part occurs during the winter. Five stations located on the windward slope of the Olympic Mountains have an average mean annual rainfall of over 100 inches, while as much as 161 inches have been recorded in a single year. As the elevations of these stations range between 300 and 700 feet above sea level, a much greater precipitation probably occurs at higher altitudes. In the Coast Range of Oregon this rainfall is somewhat less, with an average mean annual total for eight stations of 87 inches. The greatest recorded occurs at Glenora in the northern part, with a mean annual fall of 130 inches. The highest annual rainfall recorded at this station is 167 inches.

The average of the annual temperatures for six stations located in the Olympics and Oregon Coast

[4] Thornthwaite's classification is quantitative and attempts to determine the critical climatic limits significant to the distribution of vegetation. It is based upon precipitation effectiveness, temperature efficiency, and seasonal distribution of effective precipitation.

Range is about 50°, or about the same as for the coastal strip. The lesser influence of the ocean, however, is reflected by an average daily maximum temperature of 60.4 and an average daily minimum of 40.6 a range of 7° more than for the coastal strip. The growing season is shorter and a greater amount of snowfall occurs than in the latter province. The Olympics and Coast Range of Oregon are included in Thornthwaite's climatic province, designated as "wet, microthermal, with adequate precipitation at all seasons."

WILLAMETTE–PUGET LOWLAND

In the Willamette-Puget Lowland the first appreciable deviation from the marine west coast climate is to be noted. The Puget Sound region lies partly in the rain shadow of the Olympics, and the Willamette Valley to a lesser degree in that of the Oregon Coast Range. These locations result in less precipitation and a wider range between the average daily maximum and minimum temperatures, as well as other continental climatic characteristics. There is considerable variation in the province, especially with respect to precipitation. The greatest influence of the rain shadow cast by the Olympics is on the west shore of Puget Sound where the mean annual precipitation is only about 18 inches. The highest annual rainfall in the entire province, that of about 50 inches, occurs at Olympia at the southern tip of Puget Sound, while the lowest on the east side of Puget Sound is about 27 inches, at Anacortes which lies in the rain shadow of the mountain range on Vancouver Island. The average of the mean annual precipitation for thirteen stations lying between the Canadian boundary and the Columbia River is slightly more than 41 inches. At most stations between 25 and 30 per cent of the precipitation occurs in the months of May to October inclusive. In small localized areas, where the soil is very porous, the humidity low, and the evaporation coefficient during the growing season high, precipitation is inadequate for optimum plant growth. The average of the annual temperature for twelve stations is 50.6°, the average daily maximum is 59.7°, and the average daily minimum is 41.8°. The mean annual temperature is therefore similar to that of the coastal strip and the Coast Range, the range between the average maximum and average mimimum being a little less than that of the Coast Range and somewhat more than that of the coastal strip. As all of the Puget Sound stations are at less than 200 feet elevation, low altitude is partly responsible for a lower range of temperature than that of the Coast Range, and Puget Sound itself has a moderating effect upon temperature.

The average of the mean annual precipitation of 6 well distributed stations in the Willamette Valley is about 40 inches. The variation is not as great as in the Puget Sound region. The lowest precipitation is 36 inches at Salem, located midway in the valley, and the highest is over 48 inches at Newberg, located at a higher altitude and nearer the foothills of the Coast Range. A slight increase in altitude to the east or west results in a slightly higher precipitation. At five stations only 20 to 26 per cent of the total falls during May to October inclusive, causing the summers here to be slightly drier than in the Puget Sound region. The average of the mean annual temperatures for six stations is 52.3°, the average daily maximum 62.1°, and the average daily minimum, 42.6°. These averages are slightly higher and the range greater than those of the Puget Sound region, suggesting the influence of lower latitude and the absence of a moderating body of water comparable to Puget Sound. The marine climate of the Willamette Valley is somewhat modified by the air currents flowing down the Columbia Gorge from eastern Washington and Oregon, bringing for short periods hot weather in summer and cold weather in winter. The southern part of the Puget Lowland is likewise affected. The annual snowfall ranges generally from 5 to 15 inches, and the length of the growing season varies from 180 to 260 days. Most of the Puget Sound Lowland is designated as having a "humid, microthermal climate, with adequate precipitation at all seasons" (Thornthwaite, 1931). The climate of a small area lying immediately to the southeast of Puget Sound and of most of the Willamette Valley is classified as "humid, microthermal, with inadequate summer precipitation." This is reflected by the vegetation of these areas, as will be discussed later. The San Juan Islands, lying between Vancouver Island and the Washington mainland, have a slightly drier climate than that of most of the Puget Sound region, and are probably to be included in the latter classification also.

CASCADE RANGE

As the trajectory of air flow continues eastward and is forced upward to pass over the Cascade Range, increased orographic precipitation on the west slope results. The increase probably continues to an altitude of about 5,000 feet or to the level of maximum cloudiness, and then as the winds descend the leeward slope precipitation declines rapidly out across the Columbia Basin. The influence of the Pacific Ocean is still present on the western slope, and a modified marine climate persists, whereas on the east slope a much greater degree of continentality prevails. Rugged topography and great range in elevation provide this mountain province with wide variation in climate. On both slopes elevation is the chief control for both precipitation and temperature. The average of the mean annual precipitation of six stations in Washington situated on the west slope under the influence of rising air currents, and ranging in elevation from 1,150

to 5,550 feet, is 101 inches. On the west slope of the Oregon Cascades, the average mean annual rainfall of seven stations is about 91 inches, with a range of 63 to 124 inches. The elevations of these stations range from 700 to 3,900 feet, and probably the average for stations at higher elevations would be greater. The failure of the storm paths of greatest and most frequent precipitation to move as far south as Oregon at certain seasons is perhaps partly responsible for this lower precipitation. Temperature records for the Cascade province are scanty for wide ranges of altitude and too localized to present reliable comparative data. For the few stations on the west slope, however, the range between the maximum and minimum temperatures average is about 25°, somewhat higher than that of the climatic provinces to the west. Heavy snowfall shortens the growing season, especially for the lesser vegetation. A greater tendency toward continentality is indicated by the fall of a higher proportion of the total precipitation during the warmer half of the year, with from 25 to 35 per cent falling from May to October inclusive.

On the east slope of the Cascade Range the climate abruptly assumes a rather continental aspect. The precipitation rapidly decreases down the leeward slopes. In Washington the average mean annual rainfall for six stations ranging in altitude from 1,150 to 3,400 feet is about 48 inches, while for Oregon the average for five stations ranging in altitude from 1,750 to 6,500 is only about 32 inches. About 35 per cent occurs during the six warm months. Snowfall is less on the east slope, and the range of both diurnal and seasonal temperature is greater. The growing season ranges from 80 to 120 days.

COLUMBIA BASIN

The area included in this province for climatic considerations lies between the Okanogan Highlands of northern Washington and the northern Great Basin province of south central Oregon, and between the Cascades on the west and the Blue Mountains of Oregon and the northern Rocky Mountains of Idaho on the east (map 1). It is not entirely homogeneous with respect to climate because of differences in relief, but for the general purposes of this study it is included in a single discussion. With the Cascade Range serving as a barrier against the full impact of the marine influence of the Pacific Ocean, and only the Rockies as a protection against the continental climate of the Great Plains, much of the marine aspect of the Columbia Basin climate is absent and modified toward continental conditions. The proportion of precipitation during the warm season gradually increases eastward, the total amount decreases, and greater ranges in the diurnal and seasonal temperatures prevail. The average mean annual precipitation from the average of eleven stations in central and eastern Washington, ranging in altitude from 700 to 2,500 feet, is about 10 inches, with a range from 6 to 15 inches. In Oregon the average of the mean annual rainfall for thirteen stations ranging in altitude from 285 to 3,000 feet is about 11 inches, with a range from 7 to 13 inches. About 37 per cent of the precipitation occurs during the warm season. The chief control in the differences in the amount of precipitation in various areas seems to be elevation. There may be a difference of as much as 5 inches on windward and leeward slopes for a thousand-foot rise in altitude. The average daily maximum temperature for eight stations is 62°, and the average daily minimum is 36°, resulting in a seasonal average range of 26° as compared to only 13° for the coastal strip. The range between the January and July mean temperature averages about 45°, while at some stations it is over 50°. The high summer temperatures, the low precipitation, and the low humidity and resultant high evaporation provide a semiarid climate over much of the Columbia Basin. Thornthwaite designates most of this area as "semiarid, microthermal, with a deficiency of precipitation at all seasons." A small, lowland area in south central Washington, comprising about 2,500 square miles, has a mean annual precipitation of about 6 inches, and is the driest area in the Pacific Northwest.

OKANOGAN HIGHLANDS

The Okanogan Highlands, lying immediately north of the Columbia Basin, provide a less continental climate than the latter, largely because of their greater altitude. The eastward-moving air trajectory descends the east slope of the Cascade Range into the Okanogan Valley which borders the Okanogan Highlands on the west, and precipitation there is about the same as in the Columbia Basin. Upon rising again to pass over the Okanogan Highlands, much of the remaining moisture is dropped in this area. The average mean annual precipitation based on records for fifteen stations ranging in altitude from 1,200 to 3,500 feet is almost 17 inches. Some of these stations lie on the border of the province where the annual precipitation is as low as 14 inches, while at higher elevations it is as high as 32 inches. Considerable variation occurs locally owing to altitude and to positions on windward or leeward slopes. From 35 to 40 per cent of the precipitation falls during the warm season, and at one station as much as 60 per cent, partly of convectional origin, is recorded from May to October inclusive. Thunderstorms are of frequent occurrence in late summer. The average annual temperature for eleven stations is 46°, slightly lower than that of the Columbia Basin. The average daily maximum temperature is 59° and the average daily minimum is 33°; hence the range is relatively the same as for the Columbia Basin, but both absolute extremes are lower by almost 3 degrees. Snowfall

ranges from 15 to 60 inches, and the growing season from 80 to 150 days. The climate of the fringe west of the Okanogan Highlands is designated as "subhumid, microthermal, with a deficiency of precipitation at all seasons."[1] As precipitation increases to the east, the climate of the central part is classified as "subhumid, microthermal, with adequate precipitation at all seasons," while the easternmost part, adjoining the northern Rocky Mountain province proper, has a "humid" climate (Thornthwaite, 1931). The boundary between the Okanogan Highlands and the Columbia Basin coincides in general with that between the forested and untimbered zones.

NORTHERN IDAHO

The Okanogan Highlands grade into the northern Rocky Mountain physiographic province, where still further modification of the west coast marine climate progresses owing to greater distance from the ocean, greater influence of the interior conditions, and the generally greater altitude. The average mean annual precipitation for four stations ranging in altitude from 1,665 to 2,100 feet in northern Idaho is 25.6 inches, of which from 35 to 45 per cent falls during the warm season. The average mean annual temperature for four stations is 44°, while the average daily maximum is 57° and the average daily minimum is 32°, or a range of about 25°, similar to that of the Columbia Basin. The January average is 23° and the July average is about 65°, a seasonal range somewhat less than that of the Columbia Basin. The snowfall, which in this region averages about 73 inches at four stations, is greater than in any other province in the area except in the Cascade Range and the Olympic Mountains. The climate of northern Idaho is designated as "humid microthermal, with adequate precipitation at all seasons" (Thornthwaite, 1931). This is the same as that of the Puget Sound region. The latter has more total rainfall but a smaller proportion during the warm season, and warmer summers and greater evaporation. The total climate of the two regions is somewhat similar as indicated by the likeness of the climax forests.

BLUE MOUNTAINS

The Blue Mountain province, largely in northeastern Oregon and subordinately in adjacent parts of southeastern Washington, adjoins the Columbia Basin which lies to the west and northwest. The rain shadow cast by the Cascades is broken here by the higher elevation of the Blue Mountains, and so the precipitation on the mountains is appreciably greater than at lower altitude nearby. The average annual precipitation for twenty-one stations ranging in altitude from 2,700 to 6,250 feet is almost 24 inches. It ranges from 13 to 43 inches, depending upon the altitude, whether in a valley or upland, and whether on the windward or leeward slope. The largest intermontane valley, the Grande Ronde at an elevation of 2,700 feet, has an annual mean precipitation of almost 20 inches. On the average, 35 to 40 per cent of the precipitation, partly of convectional origin, falls during the warm season in the Blue Mountains, indicating the lessening influence of the Pacific Ocean and the tendency toward continentality. The average annual temperature for eight stations is 46.6°, the average daily maximum is 59°, and the average daily minimum is 34°. The range in annual temperatures of 25° is comparable to that of most areas east of the Cascades, with the extremes lower than that of the Columbia Basin. The range between the January and July mean is about 40°, considerably lower than that of the Columbia Basin. The annual snowfall ranges from 40 to 100 inches, and the growing season is from 100 to 160 days long. The climate of the more northern and higher part of the Blue Mountain province is classified as "subhumid, microthermal, with adequate precipitation at all seasons," while toward the south and west it grades into the semiarid climate of the Columbia Basin (Thornthwaite, 1931).

NORTHERN GREAT BASIN

The climate of the Northern Great Basin in south central Oregon does not differ much from that of the Columbia Basin to the north The average of the mean annual precipitation for eight stations ranging in altitude from 4,000 to 6,000 feet is almost 14 inches. This slightly greater precipitation than that of the Columbia Basin to the north reflects the generally greater altitude of the Great Basin province. Temperature data are scanty for this province, but the average January mean for six stations is 27.3°, the July mean is 65.5, and the average annual temperature is 45.6°. The range between the January and July mean is about 38°, or almost 10° less than that of the Columbia Basin. The slightly lower temperatures and the higher precipitation than those in the latter province place the Northern Great Basin province in a "semiarid, microthermal" climatic zone, with a deficiency of precipitation in summer rather than at all seasons.

FORESTS OF THE PACIFIC NORTHWEST

GENERAL CLASSIFICATIONS

The wide range of climate in the Pacific Northwest results in several forest complexes, each one being rather distinct in its characteristics and requirements as a biocoenose. Yet most of the species of which the forests are composed have a wide geographic range within the region and are to be found in several of the forest climaxes. They may assume a different degree of abundance and dominance and successional status in each forest, depending upon their associates, their relative ecological requirements, and the environment. Each tree species carries out its life cycle only in relation to its environment in conjunction with its inherent characteristics. Thus a given species may be climax in one area and subclimax in another, dominant in one region and subdominant in another, a pioneer invader in one locality and a member of the permanent community elsewhere, the most abundant species in one forest and uncommon in others, and a temporary member of one association and permanent in others. Its response to similar physical environments may vary owing to the influence of different associates, so that its rise and fall in each community do not necessarily portray the same trend of environmental change. For example, lodgepole pine is a pioneer invader of sand dune and bog on the Pacific Coast, soon to be replaced by Sitka spruce and western hemlock. In the pumice-covered areas in the Cascades of southern Oregon it is apparently a permanent species and has been since the eruption of Mount Mazama, responsible for the pumice mantle (Hansen, 1942c). Some species of the Pacific Northwest thrive only within a restricted range of environmental conditions, while others will flourish under a wide range of habitats. Certain species are not fitted for competition and thrive only when the permanent forests are destroyed. Then, as the climax species regain a foothold, they crowd out the original invaders and persist until they are again destroyed by fire, insects, diseases, or other agencies.

Although some of the forest tree species of the Pacific Northwest have a wide geographic range, they have their maximum development where the environment is at an optimum. This does not necessarily mean that the climate or any other one factor or group of factors is at an optimum, but that the total impact of the environment, both physical and biological, is most favorable. Competition may prevent a species' maximum development in an area of optimum climate, while absence of competition may permit its greater development in a region with less favorable climate. All groups of environmental factors, physiographic, edaphic, climatic, and biotic, form an interrelated complex, the force of each factor modifying the effect of the others. Not only may the limiting factor belong to any one group, but the others may assume the limiting role if the balance is altered. Aside from relative response of the several species to the physical environment, tolerance of shade, life span, amount and frequency of seed production, initial seed-bearing age, viability, ability of seed to survive fire, rate of growth, size, adaptability, resistance to disease, wind and fire, and depth of root system are some of the inherent characteristics that in their compromise with those of other individuals of the same and different species determine their ability to thrive under the existing environment.

The vegetation areas of the Pacific Northwest have

been delimited and classified upon several bases. These criteria include climate, growth forms, physiography, climax concepts, natural phytogeographic areas, and physiological ·behavior. Each of these classifications reflects the views and concepts of its author, and while formulated upon different data some of them are very similar in the final analysis. The differences are largely due to the use of different terms and degrees of subdivision. Baker ·(1934) shows an excellent correlation between four classifications of the vegetational units in the western United States. Of the purely climatic classifications, Thornthwaite's (1931) seems to be best for correlation with the vegetation provinces of the Pacific Northwest. This classification and its basis have been previously mentioned. Livingston and Shreve (1921) based their classification on vegetational criteria with little consideration of climatic, physiographic, and floristic factors. The greater portion of Pacific Northwest forests is classified as northwestern hygrophytic evergreen forest. Shantz and Zon (1924) classified the natural vegetation according to phytogeographic provinces. The vegetation units are described as being characterized by similar physiognomy, and are measured in terms of the vegetation itself and not in terms of temperature or other environmental factors. They assumed that the natural vegetation is a better indicator of environments than any one factor or group of factors. In the western region, owing to the abrupt changes in topography and great climatic range, the forest provinces are not continuous. Zon's classification of forests and Shantz' classification of grasslands seem to be valuable in relation to interpretation of the past vegetation and will be discussed later. Merriam's (1898) life zone concept has been found to be very useful in the classification of biotic provinces and in its practical application to studies of both flora and fauna. It seems to be more valuable and applicable to the western part of North America than to the eastern part. In the west the life zones are much better defined than in the east where there is greater climatic and physiographic uniformity, thus permitting more gradual transition from one zone to another. Merriam's life zone concept has been criticized by some workers, but it is extremely convenient to use and no one has been able to devise a better rule-of-thumb method for delimiting vegetation provinces in the Pacific Northwest. Merriam's life zones are determined by heat and rainfall as influencing the limits of migration of species in the higher latitudes and at higher elevations. He evaluated heat by the summation of mean daily temperatures above 43° F. from the time growth begins in the spring to the time that growth ceases in the fall; that is, by the remainder index. Differences in rainfall were also used in defining the zones,·but total rainfall was used

rather than a system of precipitation efficiency such as used by Thornthwaite. The life zone concept has been applied in the Pacific Northwest by only three workers, Piper (1906) for Washington; Bailey (1936) for Oregon; and Jones (1936, 1938) for the Olympic Peninsula and Mount Rainier. Merriam's life zones are applied best in regions where temperatures and rainfall are closely correlated, and they have come to be known by their vegetational aspects rather than by the temperature indices computed after Merriam's method. Correlations with other biotic provinces delimited by other methods can readily be shown in some cases. Clements' (1938) classification of vegetation provinces upon the basis of life form and climate involves the theory that the best measurement of climate is the plant community. His climax formations are determined by the highest type of vegetation that can be supported under the existing climate. They imply a long period of climatic stability in which the plants have become adjusted to one another. When disturbed, the climax replaces itself. Most of Clements' forest climaxes can be correlated with Zon's forest provinces. The forester has devised a method of classifying vegetation to meet his own needs; namely, the forest type. This classification is necessarily concerned to some extent with economic aspects of the forest; that is, the timber productivity; but it also involves certain ecological and phytogeographic considerations. The forest type classification of an area has certain advantages over the more general classifications in that it includes a mathematical estimate of the proportions of the several species present. This is valuable in pollen analysis of the sedimentary columns because it provides a statistical basis upon which to compare pollen proportions of the uppermost horizon of the pollen profiles with the proportions of the several species of forest trees in the vicinity. Maps of forest types are usually on a large scale, and different types are more sharpy defined than the more generally delimited climatic, physiographic, and phytogeographic provinces, and vegetation climaxes.

The most recent classification of biotic provinces in North America has been devised by Dice (1943). His biotic province covers a considerable and continuous geographic area and is characterized by the occurrence of one or more important ecologic associations that differ, at least in proportional area covered, from the associations of adjacent provinces. In general, Dice's biotic provinces are characterized by peculiarities of vegetation type, ecological climax, flora, fauna, climate, physiography, and soil. Undoubtedly this classification has its place, but the delimited areas are too vast and the boundaries cut across too many physiographic provinces, climatic provinces, and areas of diverse vegetation to be of much value in their application to pollen analysis and the interpretation of past vegetation and climate.

The region represented in this study includes parts of the Oregonian, Californian, Montanian, Palusian, and Artemisian biotic provinces.

FORESTS OF WESTERN OREGON AND WASHINGTON

Most of the region from an altitude of about 3,000 feet on the western slope of the Cascade Range to the Pacific Ocean is classified by Zon (1924) as the Pacific Douglas fir of the cedar-hemlock forest. This same region is designated as the cedar-hemlock climax formation of the coast forest by Clements (1938). This area also lies within the Humid Transition area, according to Merriam's life zone classification. Pacific Douglas fir comprises an area of about 54,000 square miles. It is very heterogeneous with respect to climate and physiography, and it is marked by several distinct forest communities, each with its own successional trends, largely an expression of the different climates. The postglacial forest succession in these several forest communities has differed considerably, so it seems pertinent to discuss briefly essential characteristics in regard to the principal species and their successional relationships in relation to the environment.

COASTAL STRIP

Two of the outstanding environmental features of the strip lying between the Oregon Coast Range-Olympic Mountains and the Pacific Ocean are the extreme marine climate and the sand dune zone. These two factors have had a profound influence upon the vegetation both of the present and the past. Sitka spruce and western hemlock are the chief dominants, and the forest of the coastal strip can be aptly designated as the spruce-hemlock climax (fig. 7). The former species reaches its maximum development in this region, and it ranges from Alaska to northern California. Other species of dominant status are western red cedar, lowland white fir, and Port Orford cedar. The last ranges southward from the middle of Oregon. Two other species that are abundant in the forest but are not part of the climax are lodgepole pine and Douglas fir. The former owes its persistence to the unstable edaphic conditions in the continual shifting of the sand, to some extent to fire, and to the maturation of peat bogs upon which it is usually the pioneer invader. In the dune zone adjacent to the ocean lodgepole pine is the first arboreal invader after the dunes have become somewhat stabilized by lesser vegetation (fig. 29). Along the south coast of Oregon, Port Orford cedar is also a pioneer invader of recently stabilized dune areas. Thickets of lodgepole near the ocean beach are low and rounded owing to sand-shear caused by the abrasive action of landward borne sand (fig. 28). Individuals are misshapen and resemble the Krumholz form of trees at timberline. An occasional specimen of Sitka spruce is found in the thickets of lodgepole pine, usually on the leeward side, but hemlock has not been observed in these situations (fig. 27). Both spruce and hemlock often occur on headlands well above the ocean beach where the soil has been stabilized. These may stand alone or occur in groups. As most of these specimens seem to be fairly old, they may represent individuals left from fire, lumbering, or cultivation.

Farther inland lodgepole pine assumes a tall straight form and serves as a windbreak for other species (fig. 30). This is still within the sand zone, but the edaphic conditions are somewhat stabilized and modified by several generations of vegetation, so that spruce and hemlock may thrive. Spruce is the first to invade the lodgepole pine stands and often invades the sand dune at the same time. As more humus is formed, hemlock invades the lodgepole and spruce communities, and in the meantime the intolerance of lodgepole prevents the development of its seedlings so that it is gradually crowded out by the other two species. Spruce and hemlock become progressively more abundant farther inland and lodgepole thins out. Still farther away from the ocean, Douglas fir becomes a part of the forest complex, while spruce becomes less abundant, giving way to hemlock and Douglas fir. As the summit of the Coast Range is reached, the precipitation abruptly decreases on the leeward slopes, and Douglas fir becomes predominant and remains so into the Willamette Valley, with the exception of some of the higher slopes where hemlock still predominates, and even noble fir may form almost pure stands of limited extent at the highest elevations. Southward from a point about midway on the Oregon Coast, Port Orford cedar enters the forest complex and often invades sand dunes with the lodgepole. Port Orford cedar also may be the initial arboreal invader of mature peat bogs. It is a more permanent species than lodgepole, however, and it extends into the Siskiyou Mountains where it has been an important forest tree. Although spruce and hemlock maintain their general predominance in the coastal forests southward to the California line, Douglas fir becomes more abundant, possibly because of the change in topography. From Port Orford southward the sand coastal plain is largely absent, and the bluffs and mountains rise abruptly from the sea. In the extreme southern part of the Oregon coastal strip, redwood occurs, while in the Coast Range and Siskiyou Mountains to the east grow western white pine, sugar pine, knobcone pine, white fir, weeping spruce, mountain hemlock, and incense cedar. Destruction of the climax forest permits temporary invasion of lodgepole in the sand dune zone and of Douglas fir farther inland.

Northward from the Columbia River in Washington the spruce-hemlock belt widens until in northern Washington between the ocean and the Olympic Mountains it extends for several miles inland up the

FIG. 28. Lodgepole pine a few hundred feet from high tide, showing effects of wind and sand-shear. Near Newport, Oregon.

FIG. 30. Older stand of lodgepole pine a little farther inland. Near Newport, Oregon.

FIG. 27. Windblown Sitka spruce a few feet from high tide near Newport, Oregon. Young lodgepole pine stand in the upper right.

FIG. 29. Young lodgepole pine invading stabilized dune area. Vegetation in foreground consists largely of salal (*Gaultheria shallon*).

Photo by U. S. Forest Service

FIG. 31.   Red alder with understory of Sitka spruce and western hemlock near Necanium Junction on the Oregon coast.

Photo by U. S. Forest Service

FIG. 32.   Douglas fir and red alder restocking old burn in the Oregon Coast Range.

river valleys into the mountains. It is in this region that Sitka spruce attains its maximum development on the Pacific Coast of North America. Western red cedar is not important in the coastal forests of Oregon and southern Washington, but it becomes much more abundant and attains larger proportions in the zone between the ocean and the Olympic Mountains. In this area almost pure stands of great-sized trees of this species occur. With their admixtures of spruce, these forests are spoken of as the spruce-cedar climax by Jones (1936). In general it can be said that the northern part of the coastal strip represents the maximum development of forests on the North American continent, and they express the extreme optimum of forest-growth conditions for those species concerned.

COAST RANGE AND OLYMPIC MOUNTAINS

No peat profiles have been obtained from the Coast Range of Oregon and southern Washington nor the Olympic Mountain area, so only a brief discussion of the forests of this area is necessary. The principal forest tree species of the Coast Range is Douglas fir. Although this area is included in the hemlock-cedar climax of Clements', neither of these species is abundant. In the cedar-hemlock climax of the Puget Sound region and the southern part of the Olympic Peninsula, Douglas fir is a subclimax species that has persisted as a result of fires that must have occurred periodically throughout the postglacial time. In the Coast Range, however, the climate is apparently too dry for hemlock and cedar to thrive, and Douglas fir maintains a climax status in this area (fig. 28). Fire is not necessary in order for it to persist (Munger, 1940). The relative absence of hemlock during the entire postglacial time is denoted by the low proportions of its pollen in peat profiles in the Willamette Valley, as will be discussed later. Other species in the Coast Range, locally abundant where conditions are favorable, are noble fir on some of the higher peaks, lowland white fir well distributed with local abundance, silver fir at isolated stations in the northern part of the Coast Range of Oregon, and becoming more abundant northward into Washington, western white pine at higher altitudes, but rare, and Oregon white oak at lower altitudes on the eastern slope of the Coast Range and into the Willamette Valley to the east. As few of the peaks of the Oregon Coast Range reach an elevation of 4,000 feet, this area is entirely within the Humid Transition life area, with the exception of a few peaks that extend upward into the Canadian life zone (Bailey, 1936). The Coast Range grades into the Siskiyou Mountains in southwestern Oregon, and here there is a convergence of forest tree species representing the coastal strip, the Coast Range, the Southern Cascades, and Northern California. No pollen-bearing sedimentary columns

have been obtained from this region, and the writer has little first-hand knowledge of the vegetation of this region.

The flora of the Olympic Mountains has been well described by Jones (1936) who made an intensive botanical survey of the Olympic peninsula. The spruce-hemlock forests of the coastal strip extend up the western flank of the Olympic Mountains where they are gradually replaced by the forests of the Canadian life zone which encircles this mountain range (Piper, 1906). The principal forest tree species of this zone are western white pine, western hemlock, and silver fir, and Jones calls this forest complex the *Tsuga-Abies-Pinus* climax. Other tree species of the lower Transition zone and the higher Hudsonian zone are also present. Lodgepole pine is not common in this zone, and Engelmann spruce, a characteristic tree of the Canadian zone in the Cascades, is seemingly absent from the Olympic peninsula. Douglas fir is locally abundant on sunny slopes, but is neither a climax nor a subclimax species. Hanzlik (1932) suggests that silver fir and not hemlock is a climax species on the western and southern slopes of the Olympic Mountains. He finds that in this locality on areas once covered by Douglas fir, and which have been free from fire for several centuries, the Douglas fir gradually disappears. "As the veterans of the original stand die, they are replaced by hemlock which have come up in the understory. He believes that they, being short-lived, are in turn replaced by the even more tolerant silver fir, which then holds the ground against all intruders—until fire or axe makes a clearing" (quoted from Munger, 1940).

The Hudsonian zone, lying immediately above the Canadian, is characterized largely by forests of mountain hemlock, Alaska cedar, subalpine fir, and silver fir. Whitebark pine, characteristic of timberline in the Cascades, is apparently absent from the Olympic Mountains. It is to be noted that high proportions of mountain hemlock pollen were found at certain levels in a peat profile from a bog located at lower elevations west of the Olympic Mountains about 10 miles from the ocean. Noble fir is apparently absent from the Olympic Peninsula (Jones, 1936). The occurrence of what the writer has identified as pollen of noble fir, in the Forks bog profile, suggests that this species may have existed in the past in the Olympic Mountains.

PUGET SOUND REGION

The best expression of the cedar-hemlock climax forest with Douglas fir as a subclimax species is in the Puget Sound region. This forest embraces the region from the lower east flank of the Cascade Range on the east to the lower east flank of the Olympics on the west, and for some distance south of Puget Sound in the Puget Lowland. In addition to the species mentioned above, lowland white fir, Sitka spruce, western white pine, and lodgepole pine are locally abundant. Lowland white fir is the most consistently present and seems to be one of the dominants of the climax forest. At higher elevations on the slopes of the Cascades, noble and silver fir become abundant (fig. 33). Sitka spruce occurs sparingly on swamps and floodplains, western white pine on gravelly knolls and on Sphagnum bogs, while lodgepole pine is often the pioneer invader of mature Sphagnum bogs and also occurs in large stands on gravelly prairies west of the Puget Sound near Shelton (fig. 35). Although Douglas fir is a subclimax species, it has persisted as one of the most abundant species in the Puget Sound region during postglacial time after it replaced the initial forests of lodgepole pine. The successional status of Douglas fir in the cedar-hemlock climax is perhaps best explained by Munger (1940).

Douglas fir is an aggressive and hardy tree. It produces some seed nearly every year; seeds in cones or on the ground will, in part, survive even a severe crown fire late in summer. The seedlings establish themselves rather readily in spite of many adverse factors and usually make a new forest cover within a decade, dominating both brush and other species of trees. Sometimes in the new "fire forest" hemlocks, cedars, and balsam firs are absent for a few decades, even though indigenous to the locality. More often there are occasional specimens of these species mixed with the Douglas fir. As the even-aged Douglas fir forest on an old burn continues its growth, hemlocks, cedars, and balsam first make their appearance in the understory. These tolerant species can survive where Douglas fir reproduction would not. The seed source of these invaders is sometimes a mystery when, as may happen, few parent trees are in evidence. By the time the Douglas fir are a century old, there is often a conspicuous understory of tolerant conifers. In its second century the Douglas fir stand thins out a good deal. Occasional trees die from wind, snow breakage, insects or disease, and this gives the tolerant species an opportunity for expansion. By the end of this century the invading hemlocks, etc., have definitely won a place in the main stand [fig. 34]. At 300 years the tolerant trees may outnumber the Douglas firs. The latter are mature, are no longer making significant height growth, and one by one are succumbing to casualties. No young Douglas firs take their place, for the canopy is much too dense; instead, there are hemlocks, cedars, lowland white firs, and silver firs ready to fill in any gaps. At 400 to 500 years the Douglas firs are becoming senile and are fast disappearing as the result of centuries of buffeting by the elements and disease.

Intensive field and experimental studies by Isaac (1943) on the reproductive habits of Douglas fir also show that it will not regenerate under heavy shade conditions.

Broadleaf species in the Puget Sound region are red alder, bigleaf maple, black cottonwood, and Oregon ash. The most abundant of these is alder, which often is a pioneer invader of deforested floodplain areas as well as higher ground.

Although most of the Puget Lowland of western Washington is forested with Douglas fir which persists as a subclimax and in a few areas possibly as a

climax, south of Puget Sound are open park-like areas that are known locally as prairies. Much of this area represents the outwash plain of the Vashon glaciation, and it is covered with sterile gravel. The principal species of forest trees are Oregon white oak, and Douglas fir, the former occurring in groves. In a few areas stands of lodgepole pine occur, and also an occasional western yellow pine can be found (fig. 36). Western hemlock also exists in more mesophytic spots, but it has not been as abundant during postglacial time as in the rest of the Puget Sound region. Nikiforoff (1937) describes "prairie islands" south of Tacoma and Olympia as having a definite prairie soil profile, entirely different from the rest of western Washington. These soils are dark or black in color and in most places not less than a foot thick, and evidently have not been occupied by forest for a long period. The soils universally developed on these "islands" are a typical product of the grassland environment. Nikiforoff ascribes these prairies to the inversion of the marine climate by the dry shadow cast by the Olympic Mountains to the southeast, the gravelly nature of the soil permitting rapid percolation of water, and a decrease in the biological pressure of the coastal strip toward its natural margins. Incidentally the annual precipitation in this prairie area is from 10 to 15 inches greater than in other parts of the Puget Sound region where the soil is just as gravelly, and where dense forests of hemlock and Douglas fir exist. Whereas these prairie-like areas probably owe their initial postglacial existence to the porous nature of the soil, it is suggested that their continued existence has been due to periodic burning by the Indians in order to maintain open ground for game and the production of their food plant, the Camas (*Camassia quamash*) (Jones, 1936). According to the testimony of old residents, the prairies were more extensive than they are at present. It is obvious that Douglas fir is rapidly invading the open areas as well as the groves of oak (fig. 37, 38). Western hemlock, which normally follows Douglas fir as the habitat becomes more mesophytic, has not been so abundant in the prairie region as in the Puget Sound region during the post-Pleistocene, as disclosed by pollen analysis of peat profiles. It remains a question as to whether the edaphic conditions will eventually be suitable for a climax forest of hemlock even if forest succession is permitted without interruption. This species is not common on the lower sites south to the Columbia River and into the Willamette Valley of Oregon.

WILLAMETTE VALLEY

The dry summers in the Willamette Valley, with less than 25 per cent of the annual precipitation occurring from May to October inclusive, and with little or no rainfall from the middle of June to the middle of September, have prevented development of even

Photo by U. S. Forest Service

FIG. 34. A mature forest of Douglas fir with some admixture of western hemlock, silver fir, and western red cedar in the Wind River valley, Washington. Trees range from 48 to 70 inches in diameter.

Photo by U. S. Forest Service

FIG. 33. Mature forest of western hemlock with some admixture of Douglas fir and silver fir, Wind River valley, Washington.

FIG. 36.   A western yellow pine with Douglas fir on the Tacoma prairie.

FIG. 38.   A grove of Oregon white oak on the prairies south of Puget Sound in Washington.

FIG. 35   Dense stand of lodgepole pine a few miles from Puget Sound, near Shelton, Washington.

FIG. 37.   Douglas fir invading prairie openings near Tacoma, Washington.

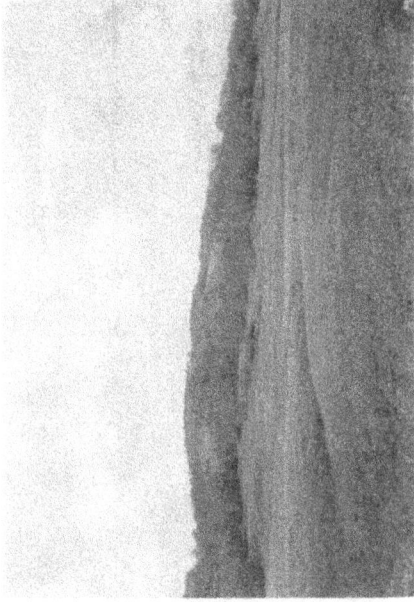

FIG. 40.   Oak on a south slope in the Willamette Valley.

FIG. 42.   Two yellow pines in the Willamette Valley.   Diameter of trunks are about three feet, excluding possibility of their having been planted by white man.

FIG. 39.   An old oak grove on the valley floor in the Willamette Valley, near Corvallis.

FIG. 41.   Remnant of a yellow pine grove in the southern part of the Willamette Valley of western Oregon, 20 miles west of Eugene.

116th Photo Section, Washington National Guard

FIG. 43. View of Liberty Lake looking north. Liberty Lake lies across the Spokane River valley from Newman Lake, and had a similar genesis.

Photo by U..S. Forest Service

FIG. 44. A Hudsonian zone forest consisting of lodgepole pine, whitebark pine, mountain hemlock, and alpine fir. Near Bird Creek Meadows, Cascades of southern Washington.

the remotest resemblance to the cedar-hemlock climax of the coast forest. Pollen profiles from this region indicate that hemlock has never been abundant during postglacial time. The principal species are Oregon white oak, Douglas fir, and lowland white fir. Although most of the forests have been removed, they were never dense and continuous like those of the Coast Range immediately to the west (Peck, 1941). Douglas fir occupies the north slopes of low hills and moister situations, while oak flourishes on the valley floor in scattered groves and on south slopes (fig. 39, 40). On the north slopes it is replaced by Douglas fir if succession is not interrupted. Lowland white fir is the other most common forest tree, and is found on moist ground, especially along water courses and on floodplains. In the southern part, western yellow pine occurs in appreciable-sized stands (fig. 41, 42). It is problematical whether these yellow pine stands are remnants of a more extensive distribution during the postglacial dry period or are of more recent invasion. Pollen analysis of a peat profile from a bog located nearby does not reveal any recorded increase in yellow pine during the warm, dry period, which is elsewhere recorded. It is not known what the successional relationship is between Douglas fir and oak. The latter is more xerophytic and is found farther down and out upon the dry areas east of the Cascades in both Oregon and Washington. At present groves of oak are being invaded by Douglas fir, and statistical studies of forests in the foothills of the Coast Range show that Douglas fir is gradually replacing the oak that thrived in areas deforested by fire and lumbering during the middle of the last century (Sprague and Hansen, 1946). It appears that the climate and the edaphic conditions of the Willamette Valley are such as to form a tension zone in which oak and fir are in equilibrium. An increase in summer rainfall would permit an increase in Douglas fir and even permit western hemlock to thrive, while a decrease in the annual precipitation would likely result in a trend toward an oak-grassland sere. Broadleaf trees in the Willamette Valley other than oak are bigleaf maple, red alder, and Oregon ash.

FORESTS OF THE CASCADE RANGE

At an elevation of about 1,500 feet in Washington and at increasing altitude in Oregon, on the west slope of the Cascades, the Humid Transition area gives way to the Canadian life zone. The upper limits of this zone are about 5,000 feet in Washington and in southern Oregon they may reach an elevation of 7,000 feet on the southwest exposures. In Oregon the Canadian zone is more or less continuous over the crest of the Cascade Range except where the higher peaks extend upward into the Hudsonian and Arctic-alpine zones. In Washington the most characteristic tree of the Canadian zone is western white pine although it is not particularly abundant. Western hem-

lock probably develops best in this zone, while silver, noble, and lowland white fir are common and attain their maximum proportions. Other arboreal species are Engelmann spruce, lodgepole pine, and western larch, all of which are more common on the east slope in the Canadian zone. Western larch ranges southward to a point about midway in the Oregon Cascades. Douglas fir is not quite as abundant on the west slope of the Cascades as in the Puget Lowland, while western red cedar is locally abundant. In Oregon, in the Canadian zone, these species are present and still others occur farther south, including incense cedar, white fir, red fir, and sugar pine, and more rarely Jeffrey pine. These species are not abundant and by no means the major members of the forest complex.

The Hudsonian zone lies above the Canadian, and in Washington it extends upward to 7,000 feet, while in Oregon it attains an elevation of over 8,000 feet. It does not form a continuous belt over the crest of the Cascades, but encircles the higher mountains. The Hudsonian zone is more extensive in Washington than in Oregon. The arboreal species of this zone are mountain hemlock, whitebark pine, alpine fir, Alaska cedar, and alpine larch (fig. 2, 44). Lodgepole pine may also be prevalent in the Hudsonian zone, especially on the pumice mantle of the southern Oregon Cascades, or elsewhere where fire or changes in edaphic conditions have temporarily eliminated the climax species from the competition. The Canadian and Hudsonian life zones are included in the spruce-fir forests of Zon (1924).

The Arid Transition timbered area forms a belt below the Canadian zone on the east slope of the Cascades, and the climate is much drier here. The principal species are western yellow pine, lodgepole pine, Douglas fir, western larch, and western juniper (fig. 45, 46). Its breadth and altitude vary with the configuration of the land, base level, and slope exposure. The annual precipitation throughout this zone ranges from about 15 to 25 inches, with an average of about 20 inches. The Canadian and upper part of the timbered Arid Transition eastward from the crest of the Cascades and for about 100 miles northward of Crater Lake lies within the area covered with a deep pumice mantle that was deposited by the eruptions of Mount Mazama and other smaller volcanoes (Williams, 1942, 1944). The pumice ranges in depth from 15 feet or more north of Crater Lake to about 6 inches in the vicinity of Bend (map 2). The pumice forms an edaphic condition that apparently has had a profound effect upon the forests. Instead of the predominance of the Canadian zone and Transition area species, much of the pumice-covered area is forested with lodgepole pine which extends in almost a continuous belt as far north as Bend (fig. 47). Interspersed with it are forests of the Canadian zone and Transition area. North of Bend western yellow pine forms a solid belt continuous almost to the

Columbia River which is contiguous to the Canadian zone on the higher slopes to the west. That the original forests of yellow pine were replaced by lodgepole after the deposition of the pumice is suggested by the pollen profiles in this region, as will be shown later. Most of the Canadian zone and timbered Arid Transition along the east flank of the Cascades is designated as yellow pine-Douglas fir forest by Zon (1924).

FORESTS OF EASTERN WASHINGTON AND NORTHERN IDAHO

The timbered Arid Transition zone, characterized largely by western yellow pine, extends across northern Washington on the Okanogan Highlands immediately north of the Columbia Basin (fig. 48). On the higher north-south mountain ranges the Canadian and even the Hudsonian zone is supported. In northeastern Washington and northern Idaho, however, western white pine and western larch become the most characteristic species and are designated as the larch-pine forest by both Zon and Clements. In this region white pine often occurs in pure stands, although the climax dominants are western red cedar, western hemlock, and lowland white fir (Huberman, 1935). White pine has been able to persist abundantly as subclimax species due to fire in the past and lumbering in more recent time. If the forest succession is uninterrupted for several centuries, the dominants obtain control, and white pine is found only as old, decadent, and diseased individuals sparsely scattered throughout the stand. In areas of repeated fire, heavy invasion of lodgepole pine and western larch occur. These two species do not usually thrive together in the same stand. The former occupies the dry knolls and exposed ridges toward the upper part of the white pine zone, while larch thrives on the north and east exposures where there is a greater abundance of soil moisture (Larsen, 1929). Both lodgepole and larch possess characteristics which permit them to invade and thrive in areas denuded by severe and recurrent fires. Lodgepole not only reproduces at an early age, but its seeds are retained in the cones for many years until a fire causes them to open and release tremendous quantities of seeds. Lack of competition then permits it to develop rapidly until more shade-tolerant species gain a foothold. Larch is long-lived and resistant to fire because of its thick bark and less inflammable foliage. Thus it will survive several fires and the large trees result in rapid restocking if the soil is not too warm and dry.

The temporary stands of lodgepole and larch are replaced by white pine and Douglas fir. These species also invade and thrive after a single fire that destroys the climax forests, or forests of white pine and Douglas fir (Larsen, 1929). After a period during which the forests are undisturbed by fire, the white pine stands mature but its seedlings fail to develop in the shade. The climax dominants in the meantime have invaded and eventually replace the pine in absence of further fire. That fire has repeatedly occurred during postglacial time is evidenced by the pollen profiles of bogs in this region, as will be shown later. Much of the white pine forests of northern Idaho lies within the Canadian life zone. At higher elevations in the Hudsonian zone occur Engelmann spruce, whitebark pine, mountain hemlock, and alpine fir. At lower elevations western yellow pine is abundant.

FORESTS OF THE BLUE MOUNTAINS

The Blue and Wallowa Mountains of northeastern Oregon support life zones from the timbered Arid Transition to and including the Arctic-alpine zone. Yellow pine is characteristic of the Arid Transition, and forms the lower timberline tree. The species of this region are somewhat similar to those of the Cascades, but noble and silver fir are apparently absent, while western hemlock is rare. That it does occur is shown by the sporadic occurrence of its pollen in peat profiles in the Blue Mountains (Hansen, 1943a). In the Wallowa Mountains limber pine (*Pinus flexilis*), a Rocky Mountain species, occurs near timberline. Other upper timberline species are whitebark pine, alpine fir, and lodgepole pine. The last species is the predominant tree in the Canadian zone, and is present in all of the timbered zones. Aspen (*Populus tremuloides*), mountain alder (*Alnus tenuifolia*), and mountain maple (*Acer Douglasii*) also are found in the Canadian and lower Hudsonian zones, while black cottonwood is prevalent along water courses in the Transition area.

VEGETATION OF NON-FORESTED AREAS OF EASTERN WASHINGTON AND OREGON

COLUMBIA BASIN

The Columbia Basin of both Oregon and Washington is largely coextensive with the Upper Sonoran life zone and timberless Arid Transition area (Piper, 1906; Bailey, 1936). This region has beeen little studied ecologically and the phytosociological units and their successional status are not well defined. This region is encircled chiefly by the yellow pine forests of the Arid Transition area. The timberless Arid Transition which forms a belt between the Upper Sonoran zone and the yellow pine forests is characterized chiefly by grasses and is generally spoken of as the Bunchgrass Prairies. There are several well defined associations present (Daubenmire, 1942). Shantz (1924) designates this area as Bunchgrass, while Livingston and Shreve (1921) call it Grassland. Clements (1938) classifies the Bunchgrass Prairie as the Palouse Prairie. It has not been possible to segregate the fossil pollen of the herbaceous and frutescent species, and in most cases even the genera, interred in

Photo by U. S. Forest Service

Fig. 46.   Pure virgin stand of western yellow pine on pumice soils near Lapine, Oregon. Pumice is from Mount Mazama.

Photo by U. S. Forest Service

Fig. 45.   Western yellow pine forest with some Douglas fir and western larch, in northeastern Washington.

Photo by U. S. Forest Service

FIG. 47. A pure stand of lodgepole pine on pumice soils near Lapine, Oregon.   Pumice from Mount Mazama, about 55 miles to the south.

Photo by U. S. Forest Service

FIG. 48.   Yellow pine and rabbitbrush near the margin of the timbered Arid Transition zone in north central Washington.

Bureau of Reclamation photo, Coulee Dam, Washington

FIG. 49. Sagebrush typical of the Upper Sonoran life zone in the Columbia Basin of eastern Washington.

Bureau of Reclamation photo, Coulee Dam, Washington

FIG. 50. A coulee in the Columbia Basin of eastern Washington, with sufficient moisture along the stream to support peachleaf willow and white alder.

Photo by U. S. Forest Service

FIG. 51. Juniper forests lying between the yellow pine zone and the timberless areas east of the Cascade Range in Oregon.

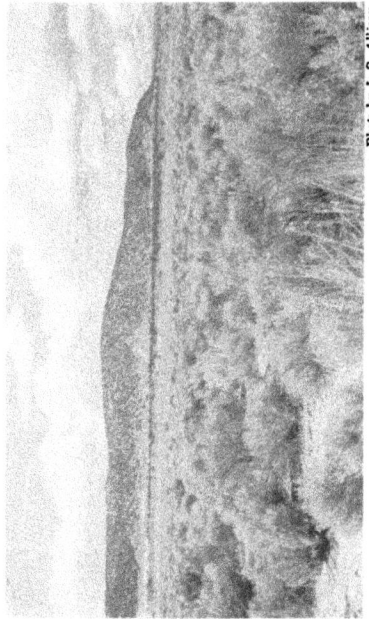

Photo by I. S. Allison

FIG. 53. Looking north from Big Hole Butte over lodgepole pine forest on Mount Mazama pumice. Darker trees in center are yellow pine. Newberry Mountain on center skyline.

Photo by I. S. Allison

FIG. 52. Silver Lake-Fort Rock basin with largely rabbitbrush and sage on valley floor, with juniper on slopes in the distance.

the sedimentary columns from this area. Only a few of the characteristic plants will be mentioned. The principal grasses include *Agropyron spicatum, Poa secunda, Festuca idahoensis*, and several species of *Bromus*. Characteristic forbs are balsam root (*Balsamorhiza sagittata*), *Achillea lanulosa*, sunflower (*Wyethia amplexicaulis*), *Sidalcea oregana, Iris missouriensis, Brodiaea douglasii, Gallardia aristata*, and various species of *Lupinus, Lomatium, Astragalus, Erigeron*, and many other forbs. Owing to intense cultivation of the area occupied by the Bunchgrass Prairie, a high percentage of the present flora has been introduced.

The Upper Sonoran zone is drier and warmer than the Transition, and assumes more of a desert aspect (fig. 49). It is classified by Zon as Sagebrush or Northern Desert Shrub; Livingston and Shreve designate it as part of the Great Basin Microphyll Desert, while Clements calls it the Sagebrush subclimax. Sagebrush (*Artemisia tridentata*) and its associates occur as climax only in the central Great Basin, which is represented in the Pacific Northwest, only in the southeastern part of Oregon. Daubenmire (1942) includes most or all of the Upper Sonoran life zone of Washington in the Artemisia-Agropyron zone, of which the loamy uplands were originally dominated by *Artemisia tridentata* and grasses, of which *Agropyron spicatum* was most characteristic. This suggests that these two species were complementary rather than competitive. It is quite probable that burning by both the aborigines and white man has played an important part in eliminating sagebrush from this zone, rather then over-grazing favoring its recent development (Daubenmire, 1942). Other characteristic shrubs of the Upper Sonoran life zone are antelope brush (*Purshia tridentata*), rabbitbrush (*Chrysothamnus nausosus* and *C. viscidiflorus*), hopsage (*Grayia spinosa*), winterfat (*Eurotia lanata*), squaw current (*Ribes cereum*), serviceberry (*Amelanchier utahensis*), horsebrush (*Tetradymia canescens*), sumach (*Rhus glabra occidentalis*), mountain mahogany (*Cercocarpus ledifolia*), and greasewood (*Sarcobatus vermiculatus*). Stiff sagebrush (*Artemisia rigidus*) is common on rocky soil and basalt outcrops. Common forbs include silver saltbush (*Atriplex argentea*), poverty weed (*Iva axillaris*), seepweed (*Suaeda depressa*), dragonsage (*Artemisia dracunculoides*), and various species of *Eriogonum, Erigeron, Penstemon*, and *Astragalus*. Next to sagebrush, the many species of *Eriogonum* are the most characteristic plants of the Upper Sonoran life zone. Russian thistle (*Salsola kali tenuifolia*) and tumbling mustard (*Sisymbrium altissimum*), introduced weeds, are abundant along graded roadsides and in sandy, fallow fields. A few of the common grasses include *Stipa comata, S. thurberiana, Oryzopsis hymenoides, Poa nevadensis, Koeleria cristata, Distichlis stricta, Elymus condensatus, Sitanion hystrix, Bromus tectorum*, and

Agropyron spicatum. Saltgrass (*Distichlis stricta*) thrives in alkaline soil; *Bromus tectorum*, introduced from Europe, is the most abundant grass of the area; while *Agroypron spicatum* is perhaps the most important native grass.

Trees are uncommon in the Upper Sonoran zone, but an occasional specimen of white alder (*Alnus rhombifolia*), peachleaf willow (*Salix amygdaloides*), and hackberry (*Celtis douglasii*) may be found along the stream courses (fig. 50).

The northeastern part of both the timberless Arid Transition and the Upper Sonoran is coextensive with the Channeled Scablands. In the coulees the Upper Sonoran vegetation extends northward and eastward into the former zone, and may also follow northward for some distance into the yellow pine zone. This is especially true in the Columbia gorge north from its junction with the Spokane River. Another significant feature of the Columbia Basin of both Washington and Oregon is the occurrence of large areas in which the soil is alkaline. These alkaline areas have developed by evaporation of playa lakes and other shallow lakes which have dried up in more recent time. Others may represent lake basins that became desiccated during the warm, dry interval of the middle Postglacial and were never reborn. In some of these dry lake basins the alkaline soil has been removed by wind or buried by windblown sand. The vegetation of the alkali flats is characteristically halophytic. Some of the principal species are saltgrass, greasewood, seepweed, winterfat, hopsage, silver saltbush, Lemmon's alkali grass (*Puccinellia lemmoni*), shadscale (*Atriplex confertifolia*), saltbush (*A. nuttallii*), orache (*A. patula*), green molly (*Kochia americana*), *Monolepsis pusilla*, and various species of *Chenopodium*. Most of these species belong to the Chenopodiaceae family, they have anemophilous pollen, and are appreciably represented in the sedimentary columns from the Columbia Basin.

NORTHERN GREAT BASIN

South central Oregon lies within the northernmost part of the Basin and Range province (map 1). This area is generally of greater elevation than that of the Columbia Basin and has a slightly greater precipitation and a wider range of temperature, both diurnal and seasonal. The soil is much poorer and in various areas is composed of dry, alkaline lake beds, volcanic ash and pumice, bare lava, sand, and swampy lands. In a few valleys there is fairly good soil, derived from silty, mucky sediments of drained swamp lands. This region has been designated as the Lake Section by Peck (1941) in his discussion of the plant regions of Oregon. Most of the Northern Great Basin of south central Oregon lies within the Upper Sonoran and Arid Transition zones, and the plants common to the Columbia Basin are also prevalent in this region.

A few plants of the Great Basin to the south barely extend over the southern boundary of Oregon. The most significant floristic features of this region are the dry lake beds and the extensive tule swamps (fig. 15). The lack of outward drainage has resulted in alkaline development of lakes and lake beds, and even the swamps become somewhat alkaline as they reach their mature stages and organic deposition ceases.

Small mountain ranges, ridges, and mesas have sufficient elevation and precipitation to support forests of the timbered Arid Transition and the Canadian zones (fig. 53). The most abundant and widely spread forest tree is western yellow pine. On the higher ridges upper slope types consisting of white fir, grand fir, Douglas fir, western white pine, western hemlock, and mountain hemlock are present, and pollen from these species occur sparsely in the sedimentary columns. Vast forests of western juniper occupy the slopes that are too dry for western yellow pine (fig. 51, 52).

SUMMARY

The vegetation of the Pacific Northwest is composed of several climaxes and many lower phytosociological units. It varies from the dense rain forests of the coastal strip to the semiarid sagebrush areas of eastern Washington and Oregon. Life zones range from the warm, wet Humid Transition and the semiarid Upper Sonoran to the Arctic-Alpine. Certain dominant species are found in several of the associations, but with different sociological status. In one association a species may be dominant, in another it may be subclimax, while in still a third it may be subdominant. The postglacial successional trends of these species in relation to the present climax in each formation considerably strengthens the interpretation of postglacial climate because these trends are consistent in response to the overall, major climatic trends. Although climate is largely responsible for the distribution and composition of the vegetational types in the Pacific Northwest, other local factors have in some areas played a more important role. The sand movement along the Pacific Coast has been a controlling factor in determining the relative abundance of the principal species in that region, while in the Cascade Range of southern Oregon the pumice has been effective in maintaining species that normally would not be dominant under the existing climate. In eastern Oregon and Washington certain areas are alkaline and support halophytic species although this is indirectly a result of climate.

In the coastal strip lodgepole pine, Sitka spruce, western red cedar, western hemlock, and Douglas fir are the principal forest tree species. The Coast Range of Oregon supports forests composed largely of Douglas fir, and, although this species is subclimax in the Puget Lowland, here it seems to be a permanent species that is not dependent upon recurring fires for its persistence. The Olympic Mountains support a number of forest types, but the chief species are Sitka spruce, western red cedar, western hemlock, western white pine, and silver fir. The most extensive development of the cedar-hemlock climax with Douglas fir persisting in abundance as a subclimax species due to fire occurs in the Puget Lowland of western Washington. The forests of this area are fairly homogeneous with only local variation from the general type. The principal arboreal species are Douglas fir, western hemlock, and western red cedar, while locally lowland white fir, Sitka spruce, western white pine, and lodgepole pine are abundant. In the southern part of the Puget Sound region prairie-like areas are common and of striking appearance within a densely forested region. In addition to many grasses and forbs, Douglas fir, western hemlock, and Oregon white oak are common. Oak first invades the grasslands, then Douglas fir, followed by hemlock, invades the oak groves, and the result is a successional trend toward the cedar-hemlock climax as the environment attains more mesic characteristics.

The vegetation of the Willamette Valley exhibits the greatest diversification from the cedar-hemlock climax. Lying between the Cascade and Coast Ranges, the dry summer climate has favored the persistence of open, prairie-like areas with Oregon white oak and Douglas fir the dominant trees. These species occur in scattered groves with much the same successional trends indicated as in the prairie regions of the southern Puget Sound region. On lower sites, lowland white fir, black cottonwood, Oregon ash, and bigleaf maple are abundant.

The Cascade Range, extending north and south throughout Oregon and Washington, supports several arboreal social units because of the wide range of climate, topography, and soil. In the Canadian life zone on the western slope the common species are western hemlock, western red cedar, Douglas fir, and silver, noble, and lowland white fir, while in the same zone on the drier, eastern slope are western white pine, western larch, Engelmann spruce, and lodgepole pine. Farther south in Oregon other species make their appearance in the Canadian zone including incense cedar, white fir, red fir, sugar pine, and Jeffrey pine. In the Hudsonian zone, which encircles the higher mountains, mountain hemlock, alpine larch, whitebark pine, alpine fir, and Alaska cedar are the arboreal species present, while in pumice areas lodgepole pine has persisted as the most abundant forest tree.

Below the Canadian zone on the eastern slope of the Cascade Range lies the timbered Arid Transition in which the most common tree is western yellow pine. Other species that are common in this zone are Douglas fir, western larch, western juniper, and lodgepole. The last species is largely confined to the pumiceous soils and recent burns in the upper part of the zone. The timbered Arid Transition zone also extends across the eastern part of northern Washington and Idaho. At higher elevations in northeastern Washington and northern Idaho, however, the climax dominants are apparently western hemlock, western red cedar, and lowland white fir, with western white pine persisting as one of the principal species owing to fires. In this respect it is similar to Douglas fir in the Puget Lowland. Other species abundant and persisting because of pyric influence are western larch and lodgepole pine, and possibly Douglas fir. In the Blue and Wallowa Mountains of northeastern Oregon and southeastern Washington the arboreal species are much the same. However, silver and noble fir are absent and there is a sparse occurrence of limber pine. Western yellow and lodgepole pine are the most abundant forest trees in these two mountain ranges.

The Columbia Basin of Oregon and Washington is generally coextensive with the timberless Arid Transition and Upper Sonoran life zones. The flora of this region consists of many species of grasses, Composites, and Chenopods, as well as many other forbs of other plant families. Also there are many species of shrubs of the last two families, the most widespread of which is sagebrush. In addition, there are a large number of plants that tolerate the alkaline soil of this region resulting from the evaporation of playa lakes each season and also the dry beds of prehistoric lakes. The vast tule swamps formed in lakes in the coulees of the Channeled Scablands of eastern Washington and the closed lake basins of south central Oregon support a large number of hydrophytic plants.

It can be readily seen that the great variation in climate, topography, and soil in the Pacific Northwest has resulted in the development of a multitude of distinct phytosociological units, involving a large number of plant species associated in a series of overlapping ecological relationships. This situation produces a complex problem in interpreting these many interrelationships into terms of climate. However, the consistency of the trends indicated in the pollen profiles, even though located in different phytogeographic and climatic regions today, probably provides more reliable data than if the vegetation and climate were more homogeneous.

CHARACTERISTICS OF PACIFIC NORTHWEST FOREST TREES

The principal trends of postglacial vegetation succession in the Pacific Northwest may be ascribed to several causes. In the glaciated region, primary succession began on the glacial till and drift and continued at a rate correlative with the rate of modification of the sterile, mineral soil and other environmental factors. In some areas normal forest succession proceeded to a stabilized climax, while in others it was interrupted or retarded by changes in the physical environment. When normal forest succession was interrupted and secondary succession was initiated, certain species were able to invade and hold the ground against the climax dominants until the disturbance factor was removed or until the conditions brought about by the disturbance were ameliorated and the environment restabilized. In certain other areas the climax was never attained, while in yet others it was never able to reassert itself after its destruction. Changes in the environment may have been gradual, such as climatic changes, or more sudden owing to fire or edaphic changes effected by volcanic activity or accelerated sand movement. The time required for reestablishment of the climax vegetation varied considerably, depending upon the magnitude, extent, persistence, and periodicity of recurrence of the disturbance factor.

While several species of forest trees may live in close association with one another, and are fairly well adjusted in their relative requirements, there is in reality intense competition between them. The degree of competition increases or decreases according to the abundance of the things they need. If there is sufficient of everything to meet their total requirements, there is perhaps no competition except for space. The limiting factors, or those that provide the basis for the keenest competition, are most critical in controlling succession. Those species that require the minimum degree of the limiting factor will perhaps best succeed. If the equilibrium is disturbed and the limiting factor is replaced by another, certain subdominant species may become dominant. The trends of postglacial forest succession in some areas have evidently resulted from alteration of the degree of impact of the limiting factor or its replacement by another. In other areas forest succession has been normal and resulted in a climax which has in general persisted to the present.

In order to interpret the postglacial forest succession as indicated by the pollen profiles, it is necessary to consider those characteristics of the principal species that permit them to take advantage of a changing environment, whether the change be gradual or abrupt. As only the major changes in the environment are perhaps depicted by the trends indicated in the pollen profiles, only the more significant characteristics need be mentioned.

DOUGLAS FIR

Douglas fir is an excellent example of a species in which the influence of its inherent traits is well reflected in its recorded postglacial trends in the Puget Sound region. Its intolerance for shade relative to that of western hemlock, western red cedar, and lowland white fir has been the principal characteristic of this species that has controlled its postglacial status in the cedar-hemlock climax. The external factor that has gone hand in hand with this trait is the result of fire. The studies and experimental evidence of Hofmann (1924), Munger (1940), Isaac (1943), and others show that, if forest succession in the Puget Sound region were continued uninterruptedly for more than five or six centuries, Douglas fir would be entirely replaced by the climax species. Apparently light has been the limiting factor in this region and, in spite of its greater seed production, lesser moisture requirements, its ability to germinate and thrive on poorer soil, as well as its greater aggressiveness and degree of general adaptability, the intolerance of its seedlings for shade eventually eliminates Douglas fir from the forest complex. Yet, in western Washington only 58 per cent of the area of old-growth virgin forest that might be expected to be of the climax type is so classed, the remainder being of the temporary type, Douglas fir (Munger, 1940). Judd (1915) stated that Douglas fir existed here only as a temporary type, which paradoxically would have vanished long ago if it had not been for the purging effects of holocaustic fires. That such fires must have occurred time and again is evidenced by its high representation throughout the pollen profiles after it had replaced the postglacial pioneer forests of lodgepole pine.

Farther south in the Willamette Valley and Coast Range of Oregon the limiting factor has apparently been moisture rather than light. Here Douglas fir has been largely predominant during the postglacial time because the area is too dry for cedar and western hemlock. Fire has not been essential for its continued dominance.

WESTERN HEMLOCK

The characteristics of western hemlock that have probably been important in regulating its successional status and trends during the postglacial times are its extreme tolerance for shade and its moisture and soil requirements. Perhaps the principal limiting factor in its distribution is moisture. It is one of the most tolerant Pacific Northwest conifers, and if plenty of moisture is available it will successfully compete with other species. It requires better soil conditions than Douglas fir, but its seedlings will thrive on sterile soil if sufficient moisture is present. It readily reproduces on the forest floor and is thus able to maintain itself continuously as long as there is no interruption by fire, permitting less tolerant species to gain a foothold temporarily. If there is insufficient moisture, other species, chiefly Douglas fir, will be able to compete successfully because light as a limiting factor is supplanted by moisture. Hemlock shows its greatest development along the Washington coast and on the middle-upper western slopes of the Cascade and Olympic Mountains where the annual precipitation is 70 inches or more and the temperature is moderate. Western red cedar has requirements similar to those of hemlock and is found in association with it, but is usually not so abundant. It apparently requires a moister soil and more humus for its germination than does hemlock.

SITKA SPRUCE

Sitka spruce is typically a fog belt species and makes its maximum development in the coastal strip of Oregon and Washington. Moisture is probably the limiting factor in its range. It is almost as tolerant as western hemlock which is its chief associate. One of its chief competitors in the invasion of the sand dune zone is lodgepole pine which is very intolerant and short-lived and thus cannot successfully compete with it after the soil conditions have become more or less stabilized. Douglas fir is also one of its chief associates but cannot survive in spruce stands where moisture is not the limiting factor. Sitka spruce apparently has been as abundant as hemlock within its range during postglacial time.

LODGEPOLE PINE

Lodgepole pine has been the chief arboreal pioneer invader in areas of postglacial primary succession, both in the wake of the retreating glacier and where new areas were formed as an indirect result of glaciation. Its consistency in invading these areas must have been due to certain characteristics which enabled it to precede the others. These characteristics probably include those that enabled lodgepole to persist close to the ice front under the rigorous conditions that must have prevailed, as well as those that encouraged its early encroachment on recently deglaciated terrain. Lodgepole pine is one of the most aggressive and hardy of the western forest trees. This is well shown by its wide geographic and altitudinal ranges. It is found at sea level from Lower California to Alaska, and inland to central Alberta, western South Dakota, and central Colorado (Munns, 1938). It reaches an altitude of 11,500 feet in the Rocky Mountains (Harlow and Harrar, 1941). Some taxonomists consider the lodgepole pine of the lower Pacific slope as a different form from that of the inland mountainous regions. They have described the mountain form as *Pinus contorta* var. *latifolia* S. Wats., while still other taxonomists consider it to be a separate species, *P. murrayana* Balf. It is not within the scope of this paper to discuss the taxonomic differences between the two forms if such exist. Undoubtedly ecotypes of this species exist as they do often for species of cosmopolitan distribution. The author has noted the stunted, twisted form at sea level in the San Juan Islands, and at elevations of less than 3,000 feet a few miles distant, the tall, straight montane form.

In both mountainous regions and along the Pacific Coast lodgepole pine responds somewhat similarly to similar changes in the environment. The characteristics that permit its pioneer invasion of new or disturbed areas are its aggressiveness and adaptability, its early seed-bearing age, sometimes at six years (Sudworth, 1908), prolific seed production, retention of viability, and wide range of soil tolerance. Cones may remain closed and attached to the trees for many years, and large quantities of viable seed are accumulated. When fire occurs, the heat causes the cones to open after the fire has passed, and the seeds find a favorable place for germination in the mineral soil, while abundant light permits the seedlings to flourish. Dense even-aged stands result, growth is slow, and maturity is attained at about 200 years. Its aggressiveness in colonizing areas of edaphic change is shown by its occurrence on the pumice mantle of the southern Oregon Cascades. The deposition of the pumice, which probably destroyed most of the forests in areas of its greatest depth, resulted in an influx of lodgepole pine in a region where the climatic climax is western yellow pine at lower elevations and spruce-fir at higher elevations. On the coast the invasion of sand dunes by lodgepole attests to its aggressive nature in colonizing areas of unstable edaphic conditions. Its adaptability and tolerance of a wide range of environment are further denoted by its early invasion of mature Sphagnum bogs on the coast, in the Puget Sound region, and in the Cascade Mountains. Thus, the maneuverability of this species probably permitted its existence under the unstable physiographic and edaphic situations of an oscillating ice-front. Other species that require a longer time before bearing seed and requiring more stabilized edaphic conditions could not compete because changes in ice position, drainage, deposition, erosion, or inundation destroyed them before seed-bearing age was reached.

Two characteristics of lodgepole pine that inhibit its permanent existence in the climax forest are its extreme intolerance of shade and its relatively short life span. After the soil and physiographic conditions become stabilized, other more tolerant species gain a foothold in the lodgepole stands and gradually crowd it out. These species, which include hemlock, cedar, spruce, Douglas fir, yellow pine, larch, and the balsam firs, are all of greater longevity, and thus outlive the original stand of lodgepole and provide seed for their continual occupancy of the site until fire or some other disturbance permits lodgepole to reestablish itself. In only a few areas in the Pacific Northwest has lodgepole been able to persist as the predominant species after its invasion of primary areas. These include the glacier-scoured rocky terrain of Mount Constitution on Orcas Island of the San Juan group and the deeper pumice mantle of the southern Oregon Cascades (Hansen, 1943b, 1942a, 1942c).

WESTERN WHITE PINE

Western white pine is widely distributed throughout the Pacific Northwest, but is only locally abundant. It ranges from the Sierra Nevada Mountains of California northward to central British Columbia and eastward into the northern Rocky Mountains of northern Idaho and northwestern Montana. It makes its maximum development in northern Idaho. White pine is primarily a montane species, typical of the Canadian life zone, but it does occur near sea level in the Puget Sound region and on the Olympic Peninsula. In the region of its maximum development it is subclimax to the cedar-hemlock-lowland white fir association, persisting largely as a result of fire. The succession following fire, however, is more complex and involves more species than that of the Douglas fir subclimax in the Puget Sound region. Its characteristics which in relation to those of its associates control its successional status are ability to grow on many types of soils, its change in degree of tolerance from seedling stage to maturity, its limited seed production which does not take place until considerable age has been reached, its germination on mineral soil, and its degree of fire-resistance. Its early invasion of deglaciated terrain with lodgepole pine, but in lesser proportions, indicates its liking for a sterile mineral soil, while its occasional pioneer invasion of bogs suggests its adaptability to soil variation. Young trees of white pine are shade tolerant, but they demand more light with increase in age. This permits germination of seeds and the development of seedlings, but more tolerant species such as cedar and hemlock inhibit the development of older trees, and the climax develops, gradually replacing white pine through loss of parent trees for continued seed production. This is somewhat the reverse of the status of Douglas fir whose seedlings cannot thrive on the forest floor. Its successional relation to western larch and the climax species is its fire-resistance relative to these species. White pine is less fire-resistant than larch and more so than western hemlock. Light, infrequent fires that destroy the climax forests encourage the persistence of white pine through provision of favorable edaphic conditions, while more severe and frequent fires destroy both the climax species and white pine, and favor the temporary predominance of western larch.

Lodgepole pine is also favored by fire in the white pine region, but because of its seed-release relationships rather than greater fire resistance. The postglacial successional trends of western white pine as interpreted from the pollen profiles reflect some of the foregoing characteristics of this species.

WESTERN LARCH

Western larch, which attains its maximum development in northern Idaho and contiguous areas of Washington, Montana, and British Columbia, is closely interdependent with white pine in its successional relationships. The principal traits which seem to be correlated with its development in relation to its associates are its intolerance of shade throughout life, best germination in a mineral soil, rapid growth of seedlings, and the fire-resistance of the older trees, the seed source for the species. Being very fire-resistant, the old parent trees with thick bark can withstand severe fires that destroy other species. Another fire occurring after seedlings and saplings of other species have developed destroys them and the lack of a seed source prevents further development until the slow process of migration furnishes seed. In the meantime, old larch trees still standing after a series of severe fires continue to seed the denuded mineral soil, and lack of competition permits the rapid colonization of areas with larch saplings. Absence of fire for a long period results in the invasion of other more tolerant species and permits white pine and Douglas' fir gradually to replace it, while continued absence of fire will allow the climax species to replace pine and Douglas fir. The occurrence of postglacial fire is well reflected by the trends of larch succession as interpreted from pollen profiles from peat bogs located within its range.

WESTERN YELLOW PINE

The range of western yellow pine or ponderosa pine east of the Cascades is generally coextensive with that of Douglas fir, and when not growing in pure stands Douglas fir is one of its chief associates. It is typically a tree of the timbered Arid Transition area, and is the farthest outlier of the forest that is contiguous with the treeless Arid Transition. It is the most xerophytic of the Pacific Northwest coniferous forest tree species. In addition to its low moisture requirements, other traits that are significant in determining its distribution and successional relationships and trends are its extreme intolerance, its susceptibility to insects, and its moderate soil requirements. Owing to its xerophytic nature, yellow pine often occurs in pure stands on the east slopes of the Cascades of Oregon and Washington. In the Oregon Cascades from Crater Lake northward to Bend, a distance of about 100 miles, much of upper zone of the area that would normally be occupied by yellow pine is forested with lodgepole because of the pumice mantle. While yellow pine thrives in fairly poor soil, it does not do well in soils of volcanic origin. It is possible that the absence of suitable mycorrhizae is responsible for its poor development on such soils. Yellow pine is also the most intolerant forest tree species of the Pacific Northwest, and requires full light after it reaches the age of twenty years. It occurs in open, park-like forests. Because of its xerophytic nature, it serves as a good indicator of past climate in some parts of the Pacific Northwest. It lies adjacent to the timberless zone, in an area of critical minimum moisture supply, and its trends reflect the advance or retreat of forests in response to increased or decreased moisture (Hansen, 1939b, 1941e, 1943d). Yellow pine is less resistant to fire only relative to western larch and Douglas fir.

It generally replaces itself after being destroyed, with no intermediate stages of succession. Its principal response to postglacial physical environmental factors has been to moisture and edaphic alteration, the latter caused by the deposition of pumice. It is very susceptible to the bark beetle (*Dendroctonus brevicomis* Lec.), and much destruction from this source has probably occurred in the past. Losses as high as 50 per cent of the timber over a period of five years have been noted (Keen, 1939). The present occurrence of yellow pine on the outwash plains southwest of Puget Sound and in the upper Willamette Valley connotes its persistence on gravelly well-drained soils and under low moisture conditions, the latter being a result of the porosity of the soil and/or low summer precipitation.

BALSAM FIRS

While one or more of the Pacific Northwest firs, lowland white, noble, silver, red, white, and alpine fir, are probably represented by their pollen at various levels in most of the sedimentary columns, they are relatively poorly represented. It has not seemed expedient to separate the species of fir upon the basis of size-range frequency because their pollen profiles seem to present little that suggests trends of succession or that can be correlated with the interpreted succession of other species. Generally they are not important in the forests of the Pacific Northwest and have probably been only locally predominant during postglacial time. As previously mentioned, silver fir is the principal climax species on the southwest slope of the Olympic Mountains (Hanzlik, 1932). In the Pacific Northwest lowland white fir has the widest distribution, occurring throughout many forest associations on the moister sites. It lies about halfway between the extremes of Pacific Northwest conifers with respect to its tolerance and moisture and soil requirements. Alpine fir occurs in the Hudsonian zone in scattered groups, and is an important timberline tree. This species probably occurred at lower elevations during the Pleistocene and early post-Pleistocene owing to mountain glaciation and colder climate. Its representation by its pollen in montane peat profiles is too slight to suggest either successional or climatic trends. White fir ranges largely in the Cascade and Siskiyou Mountains of southern Oregon, and its pollen is sparsely found in peat profiles in the vicinity of Crater Lake. It attains its maximum development in the Sierra Nevada Mountains of California where it occurs as a climax species with sugar pine or red fir. It is the most xerophytic of the western firs. Noble fir is a Canadian zone species and seldom is found in pure stands. Its pollen is present in montane peat profiles, but it has been included with that of the other firs. Red fir, the largest of the American balsam firs, ranges in the southern Cascades of Oregon and far down into California. Its pollen record is too scant to portray postglacial successional and climatic trends.

ENGELMANN SPRUCE

Engelmann spruce is a montane species, distributed largely throughout the Rocky Mountains, but it is also abundant in the Cascade Range of Oregon and Washington, the Blue Mountains of northeastern Oregon and the Okanogan Highlands of northern Washington. While it is a common tree and may occur in pure stands, it is not strongly represented in post-Pleistocene sedimentary columns. It is an important component of the spruce-fir forests of the upper slopes and is associated with most of the mesophytic Pacific Northwest conifers. It ranges from 1,500 to 12,000 feet in elevation. Its silvical requirements are similar to those of western hemlock. The pollen record for this species is too scanty to interpret its postglacial trends upon the basis of its principal characteristics.

WHITEBARK PINE

One of the principal timberline trees of the Pacific Northwest is whitebark pine. It grows under extremely rigorous conditions and does not form stands to any great extent. Its chief associates are alpine fir, Alaska cedar, and mountain hemlock in the Cascades. Although it occurs on rocky, poor soil, it thrives best on well-drained loams but apparently cannot compete with other species where conditions are more favorable for them. As an indicator of extreme climatic conditions, its increased or decreased representation in the pollen profiles suggests correlative increase and decrease in favorable climatic conditions. The identification of its pollen is uncertain because it overlaps the size range of both western white and lodgepole pines. It is undoubtedly represented to a significant degree in montane peat profiles within range of pollen dispersal to the Hudsonian life zone. The arboreal associates of whitebark pine, mountain hemlock, alpine fir, alpine larch, and Alaska cedar, have somewhat similar silvical characteristics, and their postglacial trends as interpreted from the pollen profiles would depict much the same changes in the environment. However, only mountain hemlock is appreciably represented in certain profiles, and its indicated fluctuations show no significance nor correlation with respect to those of other species.

SUGAR PINE

The largest of the American pines, the sugar pine, ranges in the Cascades of central Oregon and the Siskiyou Mountains of southwestern Oregon and southward in the Sierra Nevada Range. It rarely or never occurs in large, pure stands, and usually is scattered throughout mixed stands of yellow and Jeffrey pine, Douglas fir, red and white fir, big tree (*Sequoia washingtoniana*), and incense cedar. It is more tolerant and requires more moisture than yellow pine, and is found on cooler and moister sites. The seeds of sugar pine are an important source of food for squirrels and other rodents, and this has probably been a significant factor in its sparse occurrence. Its longevity has perhaps been responsible for the persistence of this nonaggressive species. As mentioned above, the pollen grains of sugar pine cannot feasibly be separated from those of western yellow pine, and its post-Mount Mazama record has not been differentiated from that of yellow pine.

OTHER CONIFERS

Among the other conifers that may occur within range of pollen dispersal to the sites of certain of the sediments of this study are Jeffrey pine, incense cedar, Port Orford cedar, weeping spruce, redwood, knobcone pine, western juniper, and western yew. These species cannot be considered in the interpretation of the pollen profiles because their pollen cannot be distinguished from that of other species owing to the fact that it apparently is not present or has not been preserved in adequate quantities to be of any correlative value, or that the species themselves have been of minor importance in the forest complex.

OREGON WHITE OAK

The most significant broadleaf species represented by its pollen in the sedimentary columns is Oregon white oak. It is widely distributed in the Pacific Northwest but it is only locally abundant. In general it ranges throughout the Puget-Willamette Lowland, the San Juan Islands, and into the mountains of southern Oregon. Two traits that control its distribution and serve as criteria in interpreting its pollen profiles and its successional trends are its intolerances and low moisture requirements. White oak is the most xerophytic arboreal species whose pollen record contributes to the picture of postglacial climatic trends. It is more xerophytic and intolerant than Douglas fir, its chief associate and with which its successional status is most closely related. While generally restricted to the Puget-Willamette Lowland, white oak is found on the east slope of the Cascade Range where it extends farther down into dry areas than does western yellow pine. In the Willamette Valley it occupies the warmer and drier sites, and invades openings in the more densely forested areas of the Coast Range foothills. It is eventually crowded out by the more tolerant Douglas fir, however, and occurs as old, decadent specimens in the young Douglas fir stands. On the drier and warmer south slopes of knolls in the valley, it is more successful than in the foothills. It is probable that, before the advent of the white man, fire by the Indians helped to maintain white oak, while since then continued fire, as well as

lumbering and cultivation, has aided in its persistence. On the "Tacoma prairies" the persistence of
oak is probably related to the porosity of the soil and
the low summer precipitation as well as to prehistoric
burning. Its most abundant pollen record occurs in
the organic sediments of the Willamette Valley and in
relation to the indicated trends of Douglas fir is an
important climatic indicator.

RED ALDER

The other arboreal broadleaf species most abundantly represented by its pollen in postglacial organic
sediments is red alder. This species is not an indicator
of climatic or successional trends within the forest, but
its pollen profiles denote localized changes adjacent
to the site of the sediments resulting from fire. Red
alder ranges west of the Cascades in Oregon and
Washington, also from middle California to Alaska.
It often occurs in pure stands where fire has destroyed
the conifer forests, and is usually present on floodplains with black cottonwood, willow, bigleaf maple,
and others. It is not permanent and prepares the
soil for other species. Its intolerance in later life and
its relatively short life span prevent it from successfully competing with the climax species. It is the
most abundantly represented broadleaf species in the
peat profiles, and its prolific pollen-bearing trait and its
early invasion of mature bogs is reflected by the presence of more of its pollen at many levels than all other
species combined. Other broadleaf species whose
pollen is present in the peat profiles in varying proportions are Oregon ash, bigleaf maple, black cottonwood, and several species of willow. The depicted trends of these species are only local and their
pollen records have no significance with respect to
forest succession.

SPECIES OF NON-TIMBERED AREAS

There are several associations of the Bunchgrass
Prairie and the sagebrush, as well as lower phytosociological units, controlled by climate and soil and,
since the advent of the white man, by cultivation, fire,
and grazing. The successional relationships are complex and not well worked out and the characteristics
of the individual species cannot be correlated with the
postglacial fluctuations as evidenced by the pollen
profiles for several reasons. First, the record of these
plants in the sedimentary columns is too scanty, and
second, the author has been unable to separate the
pollen of either the species or the genera of the three
most significant families presented. The predominant flora of the Bunchgrass Prairie, largely contiguous
to the timberless Arid Transition area, consists of
grasses, most of which are anemophilous. While
these grasses are probably less xerophytic than those
of the sagebrush areas or Upper Sonoran zone, no conclusions can be drawn concerning the postglacial

expansion or contraction of these areas in response to
climatic changes upon the basis of the grass pollen
fluctuations. Two other plant families whose members are important in the physiognomy of both areas
are the Compositae and Chenopodiaceae. Species of
the former are mostly entomophilous and therefore
not proportionately presented in the pollen profiles,
although some of the sages have windborne pollen
(Wodehouse, 1935). The Chenopods, however, are
anemophilous and are significantly represented in
certain pollen profiles in the Columbia Basin. As a
group, the Chenopods are indicators of more arid
conditions than the grasses, particularly those of the
Bunchgrass Prairie. Their recorded increase and
expansion suggest a drier climate in two ways. Increased aridity would be reflected by those species
that thrive under semi-arid conditions on nonsaline
soil. A drier climate would tend to increase the number and extent of saline areas by increased evaporation
of standing water, both seasonally and permanently.
The halophytic species would then expand their ranges
and perhaps be more strongly represented in the pollen
profiles. Development of the warm, dry middle postglacial period is well disclosed by the pollen profiles of
grasses, Chenopods, and Composites in sedimentary
columns from the yellow pine forests near the grassland boundary, as will be shown later.

SUMMARY

Interpretation of the postglacial vegetation history
and climate from the pollen profiles requires a critical
appreciation of the characteristics of the several species
involved, and their ecological relations to each other,
groups of others, and to the physical environment.
In the Pacific Northwest the occurrence of certain
species in several distinct phytogeographic and
climatic provinces especially requires a knowledge of
their successional relationships. The postglacial trend
of a given species in one phytogeographic province
may be quite different from that in another area, and
yet it may portray a response to similar changes in
environment. On the other hand, similarly recorded
trends of a species in two or more pollen profiles from
different climatic provinces may depict a response to
quite different environmental changes. The phytosociological status and successional relationship of
each species depend upon its inherent characteristics
that control its response to the physical and biological
environment. Those characteristics that have apparently been most important in determining postglacial succession and the cognizance of which provide
a basis for interpreting the pollen profiles are tolerance
of shade, longevity, amount and frequency of seed
production, initial seed-bearing age, resistance to
fire and insects, moisture and soil requirements, and
general aggressiveness. Although climate is perhaps
the most potent physical factor of the environment

that controls plant succession, in certain parts of the Pacific Northwest other factors have had a greater influence than climate. These factors, influencing succession through the characteristics mentioned above, are vulcanism, fire, and sand movement. Of these, fire has evidently been the most effectual and widespread, while vulcanism has been responsible for producing edaphic conditions that have been instrumental in maintaining forest types inconsistent with the climatic environment. Species that have benefited by recurring fire are Douglas fir in the Puget Lowland, western white pine, and western larch in northern Idaho, lodgepole pine in many areas, and perhaps oak in the Tacoma prairies and in the Willamette Valley. Lodgepole pine has benefited most from volcanic action, foresting vast pumice areas in the central Oregon Cascades that would normally support a climatic climax of western yellow pine. This species also has been greatly favored by periods of accelerated sand movement on the Pacific Coast, and has invaded new sand areas that either buried existing forests or destroyed them.

POSTGLACIAL VEGETATION HISTORY

CAUSES OF POSTGLACIAL FOREST SUCCESSION

Interpretation of the major and general trends of postglacial forest succession from the pollen profiles must necessarily be conditioned by the limitations and sources of error inherent in the method of pollen analysis. The following interpretations are made with consideration of the characteristics of the species concerned as expressed by their present geographic distribution, successional relationships, and apparent response to the existing environment. Minor and local changes in the environment may be reflected in the pollen profiles and tend to distort the picture of regional succession. It is believed, however, that the major recorded trends have been the result of regional changes in the environment. The minor and short-period changes are probably in most cases submerged in the major trends. The latter seem to be consistently and systematically recorded in each natural province, and are considered to be a fairly reliable record of response to regional variation in postglacial environment.

The cause of the principal trends that are perhaps most strongly emphasized in the pollen profiles include climatic changes, normal forest succession, fire, volcanic activity, and accelerated sand movement. Causes of local changes that are probably weekly reflected in the pollen profiles are insect and fungal disease, landslides, snowslides, wind, and small fires immediately adjacent to the site of accumulating sediments. It seems that forests destroyed by such agencies would be limited in area and that the secondary succession initiated would be of such small relative

consequence as to alter little the picture of the major trends in a region. These changes are sometimes depicted by fluctuations in the abundance of pollen of nonforest species, while the pollen proportions of forest tree species remain much the same. A charred horizon in the sedimentary column indicates fire on or near the bog, while above the charred layer there may be an enormous increase of a species that temporarily thrives as a result of the pyric influence. Such is the case with red alder, which usually shows a sharp increase immediately above fire horizons, in areas west of the Cascade Range. By working out the average trend for each principal species for all the profiles in a natural province, the minor and perhaps meaningless fluctuations are eliminated, and a smooth curve showing the major general trends results. The number of profiles, however, is sufficient to permit such a presentation of the general trends for only the Puget Sound region and the Columbia Basin of eastern Washington.

POSTGLACIAL FOREST SUCCESSION IN THE PUGET SOUND REGION

The pollen profiles from the Puget Sound Lowland of western Washington may well serve as a basis or pattern for the interpretation of those from other regions in the Pacific Northwest with respect to the major trends of forest succession and climate. This region is fairly homogeneous in regard to climatic, edaphic, floristic, and physiographic conditions. The sedimentary columns vary in thickness, but the typological succession is similar and the volcanic ash stratum is consistently present at about the same relative stratigraphic positions. The pollen profiles reveal much the same general trends which are logical in view of the relative characteristics of the species involved. Also a large number of sedimentary columns have been obtained from the Puget Sound region and the average is a more reliable index to postglacial trends than that based on few profiles. The occurrence of the sedimentary columns on glacial drift provides a systematic time marker as a starting point, and they are considered to be more or less the same age. The climatic trends, however, are perhaps less definitely indicated because the postglacial climate has been moderated and held somewhat static by the marine influence of Puget Sound. Profiles from the Columbia Basin where greater continentality prevails reveal more definite climatic trends. Errors in pollen identification are almost absent, however, because of the low proportions of winged conifer pollen.

LODGEPOLE PINE

As the Vashon ice melted in the Puget Sound region, the pioneer arboreal invader was lodgepole pine. This species is consistently represented as such in the lower levels of fourteen sedimentary columns that lie directly upon glacial material, regardless of their thickness, type of sediments, their geographic position, physiographic setting, or typological succession. The pollen proportions of this species range from 57 to 84 per cent at the lowest levels, with an average of 74 per cent (fig. 54). As previously mentioned, lodgepole is probably over-represented, but its recorded wide predominance suggests that it was the most abundant invader in the wake of the retreating ice. This species probably grew close to the ice front and readily adjusted its position as a forest under the unstabilized physiographic and edaphic conditions because of its characteristics as noted in a previous chapter. Absence of other tree species due to the rigorous conditions eliminated competition for light. That lodgepole thrived close to the ice is suggested by the occurrence of its pollen in the sand and even fine gravel, probably deposited when the glacier or fragments of dead ice were in the vicinity. The movement of air currents was probably off the glacier, which prevented the air transportation of lodgepole pine pollen from any great distance beyond the glacier margin to the sites of the accruing sediments.

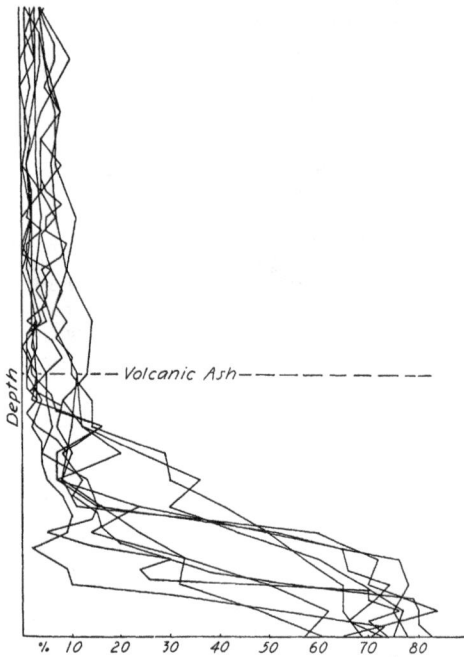

FIG. 54. Typical lodgepole pine profiles from ten Puget Sound sedimentary columns. In these and the following group profiles of a single species the thicker columns have been contracted and the shallower ones have been extended. With the exception of the volcanic ash stratum, the starting point for contraction or extension, and the surface level, it is not assumed that the horizons are synchronous.

The highest proportion attained by lodgepole pine in Puget Sound sedimentary columns is 90 per cent in the Cold Spring profile on Orcas Island, five levels from the bottom. Whereas the average of the lodgepole pine pollen profiles reveals a decrease immediately upward from the bottom, owing to adjustment of individual profiles in attaining the average, most of the individual profiles record an actual increase from the lowest to the second or third level (fig. 54). This slight rise may be due merely to the peculiarities of pollen analysis as a method, or it may represent a slight readvance of the glacier which brought about unfavorable edaphic and physiographic conditions interrupting the initial expansion of Douglas fir forests that eventually succeeded those of lodgepole. This would be somewhat analogous to a possible readvance of late-Wisconsin ice as reflected in eastern North American profiles by a slight increase in spruce after it had begun its postglacial decline (Hansen, 1937; Deevey, 1943). After this minor increase, lodgepole rapidly declined to an average of only 6 per cent at the volcanic ash horizon. The following trend is then one of more gradual decline, the lowest average proportion of 2 per cent occurring at the top (fig. 55). In some of the sedimentary columns lodgepole pollen is absent in the upper horizons.

The postglacial invasion of lodgepole pine in the Puget Sound region was probably the result of its persistence close to the ice front. Boreal or timberline species were either absent or could not compete with lodgepole under the edaphic and physiographic instability. As the ice retreated, these conditions were moderated and species such as Douglas fir and hemlock, requiring a longer period before bearing seed, gradually advanced, replacing the less tolerant and shorter-lived lodgepole. This is in accordance with the fact that of all Pacific Northwest forest trees lodgepole is the least permanent in the forest complex, persisting only where there is absence of competition or where the soil is unfavorable for other species. In only one profile, that of Mount Constitution of Orcas Island, does lodgepole maintain its initial postglacial predominance to the present (fig. 56). The terrain in areas adjacent to this profile is glacier-scoured, has very little soil present, and so has permitted lodgepole to thrive in the absence of competition. In a profile on Lulu Island, British Columbia, the normal postglacial trend of lodgepole is obscured by its invasion of the bog surface before it had been replaced by other species in adjacent upland areas. In still another profile, near Rainier, Washington, near the southern limits of glaciation in the Puget Sound lowland, lodgepole is recorded to high proportions until almost the time of the volcanic activity (fig. 57).

A time differential in the initial sedimentation in the several sedimentary columns in relation to their geographic location and progressive deglaciation is

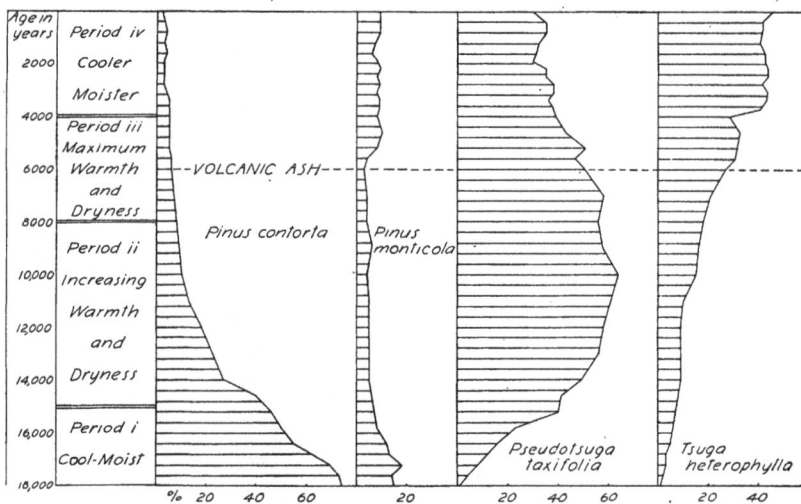

FIG. 55. Average of ten profiles from Puget Sound sedimentary columns. The volcanic ash level of all sections is considered to be of common age. Contraction and expansion of profiles begin at that level.

not reflected by the pollen profiles of lodgepole pine. The variation in thickness of the profiles beneath the volcanic ash stratum is no criterion for relative ages because of the different rates of accumulation. The thickness of the columns below the ash, recording similar trends for lodgepole, is proportionate in each profile. This suggests that about the same period of

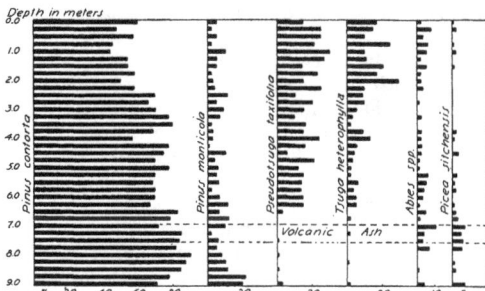

FIG. 56. Pollen diagram of Mount Constitution (Orcas Island) sedimentary column, showing predominance of lodgepole pine throughout.

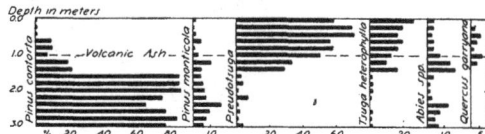

FIG. 57. Pollen diagram of Rainier sedimentary column showing persistence of lodgepole pine until almost the time of volcanic activity estimated at 6,000 years ago.

time was involved before lodgepole was replaced by other species and that there was a rapid and nearly simultaneous freeing of the Puget Lowland of ice.

WESTERN WHITE PINE

Western white pine was the next most important pioneer invader on deglaciated terrain. Its greatest proportions are recorded in the lower levels of most sedimentary columns, and then it generally declines to a minimum which is maintained to the present (fig. 58). White pine proportions range from 6 to 36 per cent at the lowest level, and like lodgepole pine it shows a slight increase immediately above the bottom in several profiles, and even the average reveals this slight increment (fig. 55). From the maximum average of 18 per cent two levels above the bottom, it declines for several levels to an average of 5 per cent, and remains constant to the volcanic ash stratum, somewhat similar to the lodgepole pine trend but on a smaller scale. The size-range of white and that of lodgepole pine pollen overlap and their similar trends immediately suggest that some of lodgepole pollen has been identified as that of white pine. The slightly greater abundance of white pine pollen above the ash, however, where pine pollen is scarce, tends to minimize this possible source of error. White pine is more abundant and widely distributed in the Puget Sound region at present than is lodgepole pine.

The recorded early postglacial abundance of white pine in the Puget Sound region is logical. It thrives in a moist, cool climate, and its soil requirements are more exacting than those of lodgepole. It was not as

DOUGLAS FIR

The postglacial pioneer forests of lodgepole and white pine were replaced by those of Douglas fir. The rate of Douglas fir expansion was probably not so rapid as indicated in the sedimentary columns because of the much slower rate of organic sedimentation in the lower levels and the same interval used in sampling the entire sedimentary column. Assuming a depositional rate twice as fast below as above the volcanic ash horizon, it required a period of perhaps 3,000–4,000 years for Douglas fir to supersede lodgepole pine (fig. 55). If it is assumed that lodgepole is over-represented in the profiles, then this figure would have to be somewhat reduced. It does not seem probable that either unfavorable climate or unmodified edaphic conditions prevented its earlier invasion of deglaciated terrain simultaneously with pine; rather its inability to persist close to the ice-front under the unstable physiographic conditions, because of its lesser maneuverability, and its older initial seed-bearing age were the chief inhibiting factors. After it had become established, its greater tolerance and longevity eliminated lodgepole as a serious competitor for the rest of postglacial time. Douglas fir is recorded to an average of 1 per cent in the lowest levels with a range of 0 to 7 per cent. It increased to its maximum average of 64 per cent some time before the volcanic eruption. Its greatest proportion of all profiles is 84 per cent at Poulsbo (fig. 59). In most profiles Douglas fir is recorded to its maximum proportion prior to the volcanic eruption responsible for the volcanic glass layer. From its peak it slightly decreases to the present, where it averages 30 per cent, with a range from 10 to 59 per cent. The curve for Douglas fir pollen proportions in most profiles has a very much saw-toothed effect (fig. 60). Most of these sharp fluctuations are probably attributable to pollen analysis as a method and do not denote changes in the forest composition. In a few cases, however, these fluctuations of greater magnitude may reflect the influence of fire. A sharp increase may record the influx of Douglas fir after fire, while a decline is suggestive of recovery of the climax dominants some time

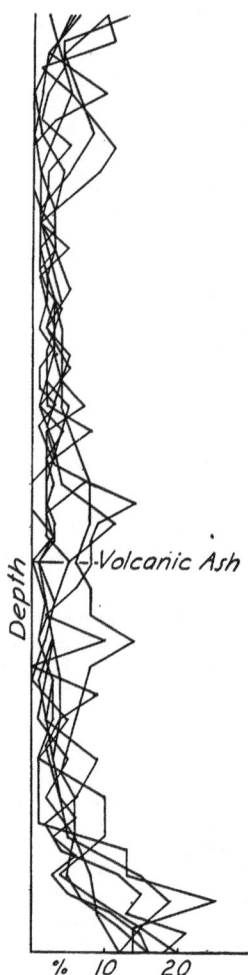

FIG. 58. Western white pine pollen profiles from seven Puget Sound sedimentary columns.

preponderant as the latter, however, because of its habit of producing seed at a much greater age. Being more tolerant and longer-lived than lodgepole, white pine has persisted in greater abundance on favorable local sites. Its decline from early postglacial time may reflect warming and drying of the climate, as the influence of the glacier became more remote. In the Poulsbo profile it shows a radical deviation from its normal record. Instead of a constant trend above the ash layer, it shows an increase in the upper levels, and attains 74 per cent at the top. Two white pine trees are present on a knoll near the center of the bog.

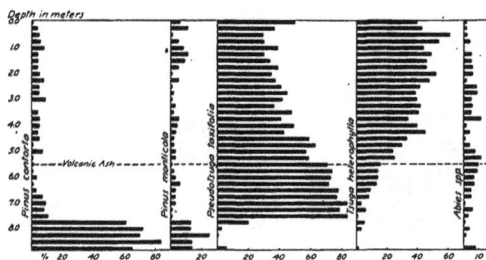

FIG. 59. Pollen diagram of Poulsbo sedimentary column, showing maximum proportion of Douglas fir in the Puget Sound.

after the fire. As the fire and resultant fluctuations undoubtedly occurred at different times in the several areas, it seems unlikely that the trends as shown by the average curve can be thus interpreted. The interval of sampling for most profiles is one-quarter meter, representing about 1,000 years if the estimated rate of sedimentation is applied. If the climax forest of the Puget Sound region, composed largely of hemlock and cedar, can be developed in five or six centuries, then complete destruction of the climax forest by fire, maturation of the Douglas fir subclimax, and attainment of a new climax may all take place during the time represented by the sampling interval. If this occurred, then a series of levels of the profile might reflect a continuous climax or subclimax, while at other points in the profile culmination of both climax or subclimax development would be omitted.

In a few of the sedimentary columns of the Puget Sound region Douglas fir did not follow the typical trends. In the Cold Spring bog on Orcas Island absence of favorable edaphic conditions has apparently prevented its replacement of lodgepole pine. In the Killebrew bog on the same island, it maintained its predominance to the present, as the summers were

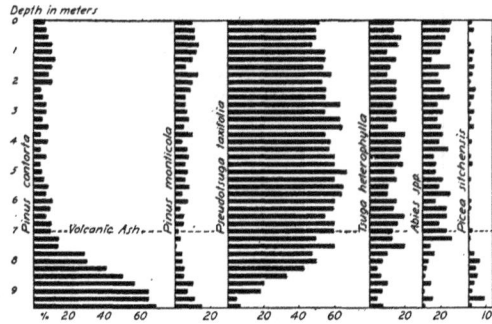

FIG. 61. Pollen diagram of Killebrew (Orcas Island) bog, showing predominance of Douglas fir to the present, probably because of the dry summers in the San Juan Islands.

probably too dry for its replacement by hemlock (fig. 61). On the gravelly prairies south of Puget Sound sedimentary sections from near Tacoma, Olympia, Rainier, and Tenino also reveal similar trends. In this region the porosity of the soil, the summer deficiency in precipitation, and perhaps the repeated burning by the aborigines have prevented western hemlock from superseding the Douglas fir at any time during the postglacial. In turn, an abundance of grass and composite pollen concurrent with the Douglas fir maximum prior to the deposition of the volcanic ash suggests a drying and warming of the climate that caused Douglas fir to retreat from the prairie-like areas. This dry period is also well recorded in certain profiles east of the Cascade Mountains and in other areas, and thus lends support to the occurrence of a postglacial warm, dry period here as well. The occurrence of appreciable proportions of oak pollen in some of the profiles on the "prairies" during this same represented period further suggests desiccation of the climate.

WESTERN HEMLOCK

The successional relationships of western hemlock and Douglas fir are well borne out by their postglacial trends. Both increased from the bottom upward, but western hemlock at a much slower rate. Although Douglas fir superseded lodgepole low in the profiles, hemlock did not do so until much higher, or possibly 4,000 years later (fig. 55). Hemlock is recorded to an average of 1 per cent at the bottom and then slowly but consistently increases upward to 27 per cent at the volcanic ash level. A more abrupt increase is recorded just before the ash level, and continues so almost to the top. The maximum average of 46 per cent is recorded about midway between the ash level and the top. In all individual profiles, the maximum proportion of hemlock is also attained above the ash stratum (fig. 62).

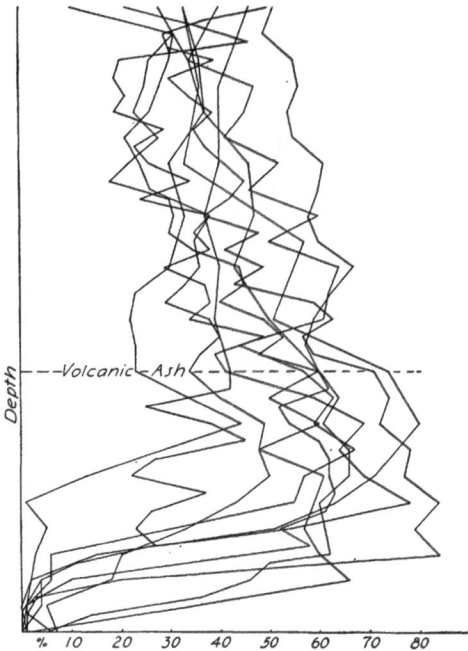

FIG. 60. Typical Douglas fir profiles from nine Puget Sound region sedimentary columns. Significant features are the rapid replacement of lodgepole pine, and the occurrence of the maximum in each profile before the time of recorded volcanic activity.

profiles this same trend occurs, with the exception of those from drier areas as pointed out above (fig. 64).

The relative trends of lodgepole pine, Douglas fir, and hemlock in the Puget Sound postglacial forest succession indicate that the pioneer forests were replaced by Douglas fir as soon as it could migrate into the area. Probably no modification of the climate or soil was necessary. Hemlock was slower to invade and expand, because of its more exacting soil requirements and its lesser aggressiveness. After it did gain a strong foothold, it was slow to expand and assume predominance because of the continued drying and warming that reached a maximum about the time of the recorded volcanic activity. Repeated fire was probably also a retarding factor. With the return of moister conditions, hemlock expanded more abruptly to supersede Douglas fir, and even fire, if it occurred, did not prevent it from maintaining slight predominance to the present.

BALSAM FIRS

The pollen profiles of the several species of true fir, considered collectively, are not particularly significant and reveal little in the way of forest succession and

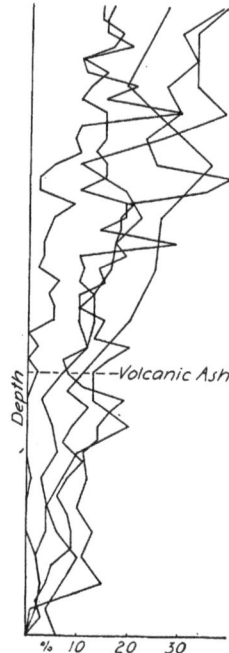

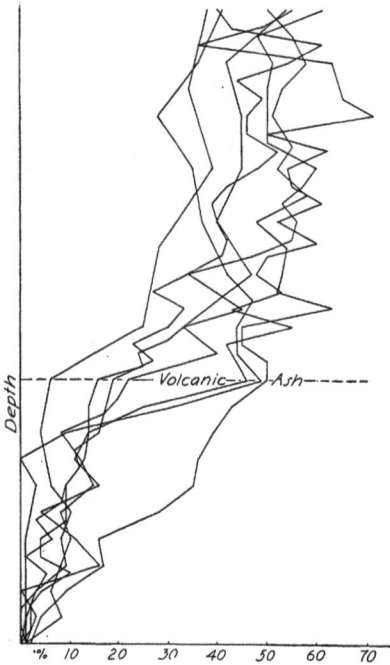

FIG. 62. Typical western hemlock profiles from seven Puget Sound sedimentary columns, showing a gradual expansion of this species below the volcanic ash stratum and its more rapid rise to predominance above.

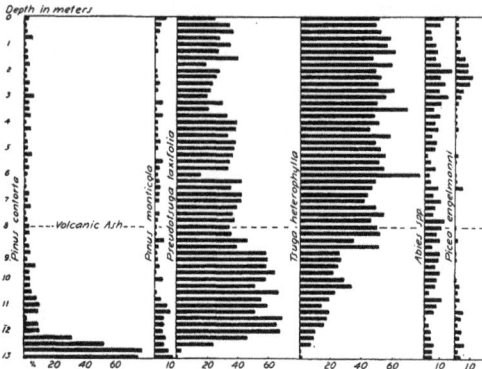

FIG. 63. Pollen diagram of the Bellingham sedimentary column. The recorded postglacial forest succession is typical for the Puget Sound region.

The greatest proportion attained is 77 per cent in the Bellingham profile (fig. 63). From the ash horizon to the top the fluctuations of hemlock and Douglas fir are largely reciprocal. A few levels above the ash layer, hemlock supersedes Douglas fir and maintains its predominance to the top. In most of the individual

FIG. 64. Pollen profiles of western hemlock of sedimentary columns located in the drier and gravelly areas of the Puget Sound region. In these areas western hemlock was unable to supersede Douglas fir after the volcanic activity due to the dry summers and/or the gravelly glacial drift.

climate. This is because of their low representation, indicating that they have played a minor role in postglacial forest succession in the Puget Sound region, and also because their pollen has not been segregated as to species. However, most of the fir pollen present is apparently that of lowland white fir which is widely distributed in the Pacific Northwest and thrives under a variety of local conditions. Its proximity to bogs and its occurrence on floodplains in greater abundance than in the forest complex perhaps tend to overrepresent it in the sedimentary columns. In general, lowland white fir is more greatly represented above the volcanic ash horizon than below although the difference is too slight to translate it into terms of climate. Apparently the environment has been somewhat more favorable after the warm, dry interval. Theoretically, noble, silver, and alpine fir should be more highly represented in the lower levels, when

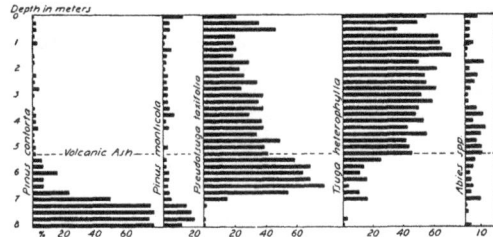

FIG. 67. Pollen diagram of the Granite Falls sedimentary column. The forest succession is typical of the Puget Sound region, with western hemlock assuming predominance after the warm, dry interval which is only slightly reflected in the forest succession.

the climate was perhaps cooler and they had not yet migrated to greater altitudes in the wake of the retreating mountain glaciers. Their pollen, however, is sparsely and sporadically scattered throughout the sedimentary columns and there is no such indicated trend. The average profile of all fir pollen is generally static and fluctuates between 4 and 10 per cent. The maximum of 33 per cent occurs in the Olympia profile (fig. 65). (See figs. 66, 67 for additional Puget Lowland profiles.)

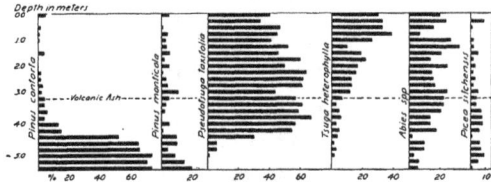

FIG. 65. Pollen diagram of the Olympia sedimentary column. The influence of the gravelly soil is reflected by the failure of western hemlock to supersede Douglas fir above the volcanic ash stratum. High proportions of lowland white fir throughout are due to local abundance of this species adjacent to the bog.

OTHER SPECIES

Two species of spruce are represented in sedimentary columns in the Puget Sound region. In the bogs near tidewater Sitka spruce pollen is abundant while those located nearer the foothills of the Cascades contain largely pollen of Engelmann spruce. In some profiles Sitka spruce may be over-represented because of its occurrence on the bog surface. The strongest record of Sitka spruce is in the New Westminster profile where it attains a maximum of 57 per cent below the ash stratum and during the dry period. In general, the scarcity of spruce pollen and its sporadic occurrence in the sedimentary columns provide little or no basis for climatic or successional interpretations in the Puget Sound region. Neither Sitka nor Engelmann spruce has played an important role in the postglacial forest succession of this region.

It is unfortunate that western red cedar, an important species of the climax forest, is so sparsely represented. Even in bogs with an abundance of cedar on their surface its pollen is conspicuously absent. The present successional status of western red cedar in Puget Sound forests is not well understood, so it is unfortunate that its paleic record contributes nothing to this problem. The highest proportion recorded is 14 per cent near the surface in the Ronald profile, near Seattle.

Red alder is the most abundantly and consistently recorded broadleaf tree in the Puget Sound region. Its pollen is usually present throughout the profiles

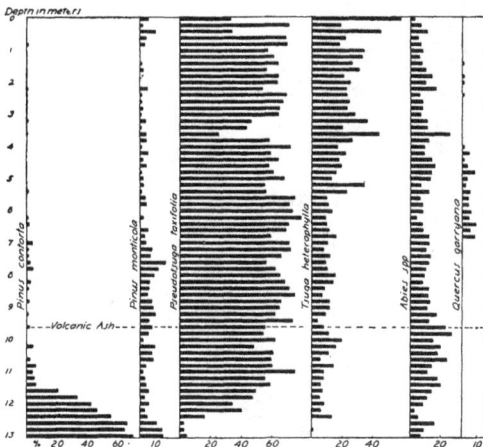

FIG. 66. Pollen diagram of the Tenino sedimentary column. The inability of western hemlock to attain predominance in the upper levels is the result of gravelly terrain and low summer precipitation. The warm, dry stage is more strongly reflected than in profiles farther north by a slight influx of Oregon white oak.

at every level, and over 300 grains have been counted at certain horizons. Its increase is often correlated with charred layers immediately below, suggesting the temporary advantage gained as the result of fire. These high proportions are of short duration, indicating rapid recovery of the forest trees. The occurrence of alder pollen also denotes its invasion of the bog surface, or its occurrence on damp sites immediately adjacent to the bog. It often is abundant in the marginal ditch that borders many Puget Sound bogs.

POSTGLACIAL FOREST SUCCESSION IN THE WILLAMETTE VALLEY

Five sedimentary columns from the Willamette Valley of western Oregon record a trend of postglacial forest succession somewhat different from those in the Puget Sound region. Although the Willamette Valley is included within the cedar-hemlock climax, both of these species are uncommon at the present and evidently have never been abundant during postglacial time. At present, the annual precipitation is as great

as or greater than that in most parts of the Puget Sound region, but the summers are too dry for hemlock to thrive. As previously mentioned, the Willamette Valley was inundated during the close of the last Wisconsin glacial epoch, and it seems probable that the sedimentary columns represent most of postglacial time.

LODGEPOLE PINE

In the profiles from Onion Flats and Lake Labish the trends of lodgepole pine are similar to those of the Puget Sound region and other parts of the Pacific Northwest. It was the pioneer invader as the glacial floodwaters subsided and permitted forests to enter on the valley floor. At Onion Flats lodgepole was predominant for more than two-thirds of the profile, while the Lake Labish profile records its predominance only in the lower third (figs. 68,69). This disparity in the relative thickness of the profiles holding the record of lodgepole pine predominance is explained by the type of sediment. Onion Flats silts and clays are much thicker and these sediments have evidently been

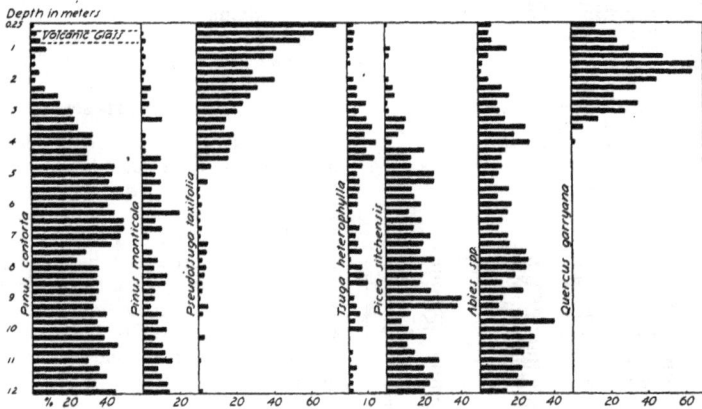

FIG. 68. Pollen diagram of the Onion Flats sedimentary column. The influence of the warm, dry middle Postglacial is shown by the influx of oak in the upper levels. An unknown thickness of sediments has been removed by drainage, cultivation, and deflation in more recent times.

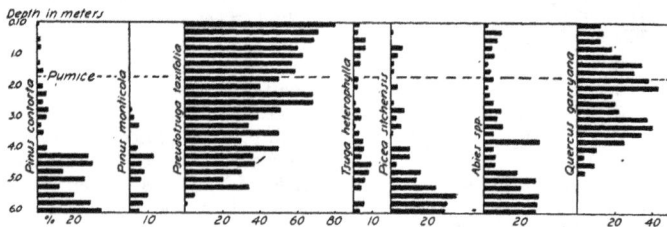

FIG. 69. Pollen diagram of the Lake Labish sedimentary column. The lower stratigraphic position of the pumice stratum than that at Onion Flats suggests that less of the upper sediments have been removed by deflation and cultivation.

of rapid deposition. The 12-meter sedimentary column here is probably equivalent in time of deposition to that of the 6-meter Labish columns. In the upper levels lodgepole diminishes rapidly, and is recorded to only a few per cent to the uppermost level. The sedimentary columns from the three sites have been truncated by subsidence and deflation, and it is impossible to estimate the thickness that has been lost. There is no record of lodgepole pine in the Willamette Valley since the advent of the white man, so the missing portion of the sedimentary columns may represent sufficient time to result in elimination of lodgepole from the forest complex. Lodgepole pine seemingly persisted for a longer time in the Willamette Valley than in most parts of the Puget Sound region. The transportation of lodgepole in the coast region today and in recent time, however, and the absence of its pollen in Willamette Valley sediments of more recent deposition tend to eliminate the coast as a source of its pollen. Transportation by stream from the Cascade Range to the east is possible, as the Willamette River heads in the mountains. During early postglacial time the streams may have been flooded at certain times, resulting in overflow and backing into tributary sloughs, with consequent deposition of lodgepole pine pollen. Its pioneer invasion of deglaciated terrain over most of the Pacific Northwest, owing to lack of competition, however, lends support to its existence in abundance in the Willamette Valley during early postglacial time.

In two other profiles in the Willamette Valley lodgepole pine did not play such a conspicuous role. In the vicinity of the Noti profile, located in the foothills of the Coast Range, lodgepole has existed in limited quantities during postglacial time (fig. 70). Its greatest abundance is recorded at the same level at which silt and volcanic glass are abundant. Its low proportions in the lower levels suggest that either the profile does not represent early postglacial time or the forests of the Coast Range which persisted through the glacial period did not contain much lodgepole pine.

In another sedimentary column near Scotts Mills in the foothills of the Cascade Range lodgepole pine was abundant in the lowest level but not predominant

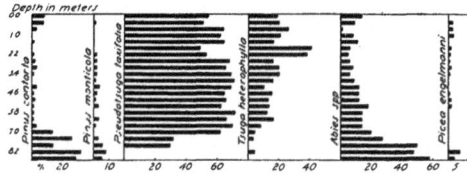

FIG. 71. Pollen diagram of the Scotts Mills sedimentary column, showing predominance of fir in the lower levels rather than that of lodgepole pine. All of postglacial time may not be represented by the section.

(fig. 71). Here, two or more species of fir were predominant. The absence of initial lodgepole predominance suggests that the sedimentary column is too young to represent the early postglacial period. Its decline upward in the profile and its representation in small proportions to the top is similar to that of the Lake Labish and Onion Flat profiles.

DOUGLAS FIR

Douglas fir expanded rapidly, apparently at the expense of lodgepole, after its initial appearance. In the Onion Flats profile it is recorded to low proportions for many levels upward, and then increases to the uppermost level. In the Lake Labish profiles Douglas fir expands rapidly almost from the bottom, and soon replaces lodgepole pine, to remain generally predominant to the top. This is similar to its trends in the Puget Sound region, and probably a result of the modification of the environment.

In the Noti profile Douglas fir is recorded irregularly below the volcanic glass layer, fluctuating contrarily to hemlock, fir, white pine, and yellow pine. Above the volcanic ash it is recorded to high proportions to the top, indicating that the climate has been too dry for hemlock to supersede it even slightly, as it did in the Puget Sound region.

In the Scotts Mills profile Douglas fir superseded lodgepole and fir low in the sedimentary column, and remained predominant to the present. At one time hemlock increased abruptly at the expense of Douglas fir, but did not supersede it.

WESTERN HEMLOCK

In the Onion Flats and Lake Labish sedimentary columns, hemlock is recorded in negligible proportions throughout, its maximum being 15 per cent. Its lowest proportions occur during the period of oak maximum, indicating its response to the warm, dry period. It reveals a slight expansion in more recent time, as the climate became cooler and moister. In the Noti profile the proximity of the Coast Range with its moister climate is reflected by higher proportions of hemlock than in the valley proper. Here hemlock is recorded to 42 per cent and does not drop below 13

FIG. 70. Pollen diagram of the Noti sedimentary column. Significant features include low proportions of lodgepole pine, higher representation of western hemlock than at Lake Labish and Onion Flats, and appreciable amount of western yellow pine pollen in the lower half.

per cent. The trend of this species does not provide additional evidence for the occurrence of the dry period.

In the Scotts Mills profile hemlock is also recorded to greater proportions than in the valley profiles. It increased gradually from the bottom to 2.6 meters, and then sharply increased to 41 per cent at 1.8 meters, its maximum of the profile. It then diminished sharply and remained more or less constant to the present, perhaps a result of fire. A period of constant proportions after its initial increase may reflect its arrested expansion during the dry period. Its decrease from its maximum to the top, however, is not consistent with the increase in moisture since the dry period. The postglacial trends of hemlock in the Willamette Valley reflect the dry summers, which have apparently been characteristic of the climate during the entire postglacial period.

ORGEON WHITE OAK

Perhaps the most significant trend is that of Oregon white oak. Whereas the influx of Douglas fir indicates normal forest succession as the early postglacial conditions were ameliorated, and the low proportions of hemlock indicate dry summers, these trends do not reveal any evidence for the dry period. In the Onion Flats and Lake Labish profiles white oak is recorded as having been absent for at least the first half of postglacial time (figs. 68, 69). After making its initial appearance, it expanded rapidly at the expense of the moisture-loving species, as well as of Douglas fir. Oak is more xerophytic than any of the northwest conifers, and its expansion during this time undoubtedly was due to the influence of the dry period, so strongly evinced in other profiles. In the Onion Flats and one of the Lake Labish profiles it superseded Douglas fir in spite of its much lesser pollen-producing capacity. The maximum of oak occurs nearer the top of the profile at Onion Flats, but the position of the pumice nearer the surface indicates a greater loss of surface sediments than at Lake Labish. In all three profiles the oak maximum occurs below the pumice layer, indicating the synchroneity of the warm, dry interval. Oak declined from its maximum and is recorded in moderate proportions at the uppermost level, consistent with increased moisture during the last 4,000 years. In the other two profiles, oak is recorded only at one or two levels and in low proportions. It evidently was never abundant in the Coast or Cascade Range forests, because of the greater precipitation over these areas.

SPRUCE AND FIR

Sitka spruce and lowland white fir are the most abundantly recorded of the other conifers. Sitka spruce is absent in the valley today and was most abundantly recorded before the advent of the dry period. It recovered slightly as moister conditions prevailed in more recent time. In the other two profiles fir is much more abundant. In the Noti profile lowland white fir is recorded most abundantly in the lower half of the profile. In the Scotts Mills profile most of the fir pollen is probably that of noble fir, and it is the predominant species in the lower three levels. The proximity of the Cascade Range with its moister climate is reflected by appreciable proportions of fir throughout the profile.

While the oak maximum is strong evidence that the dry period was felt in the Willamette Valley, the corroborating evidence of grass influx is seemingly absent. According to hearsay, great fires occurred when the white man first entered the valley in large numbers from 1845 to 1855. Prior to settlement by the white man the Indians are said to have kept the country burned off, as on the Tacoma prairies. In recent time Douglas fir has invaded the grassland and oak groves, indicating the natural climax (Sprague and Hansen, 1946). The absence of the upper sediments of the Onion Flats and Lake Labish profiles is unfortunate as they might have held a record of grass influx due to burning by both Indians and white man in the last several hundred years.

POSTGLACIAL VEGETATION OF EASTERN WASHINGTON

GENERAL CONSIDERATIONS

The pollen profiles most significant as to the postglacial climatic trends are those from the Upper Sonoran and Arid Transition zones of eastern Washington. Those that lie within and near the perimeter of the yellow pine zone or timbered Arid Transition hold the most definite evidence for these climatic fluctuations. The yellow pine forests, occupying the most arid of the timbered regions of the Pacific Northwest, adjacent to the Bunchgrass Prairie, have been influenced more by climatic trends than have the more mesophytic forests because the precipitation is and probably has been at a critical minimum. A slight decrease would result in a contraction of the yellow pine forests, while a moderate increase in precipitation would undoubtedly permit their expansion. Such changes in climate, however, would perhaps be more strongly reflected by a corresponding adjustment in the position and extent of the grassland and sagebrush areas because of their more limited representation in the sedimentary columns. Yellow pine, shedding more pollen than grass and sagebrush, and with its pollen being farther dispersed, probably has been consistently over-represented in the pollen profiles in relation to the grasses, Chenopods, and Composites. This degree of over-representation is further amplified in some of the sedimentary columns by the position of the pine forests upstream, permitting water transportation of pollen to the sites of those

Depth in meters

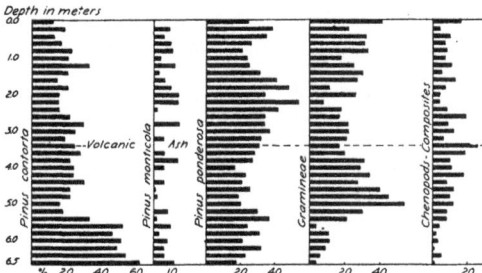

FIG. 72. Pollen diagram of the Harrington sedimentary column from the Channeled Scablands of Washington. Although 20 miles from the nearest forests, lodgepole and western yellow pine are appreciably represented at the top.

Depth in meters

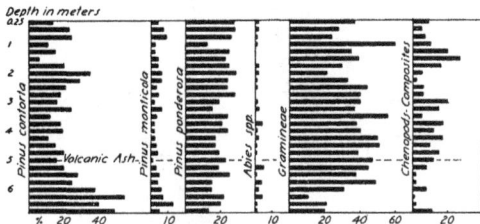

FIG. 73. Pollen diagram of the Crab Lake sedimentary column from the Channeled Scablands of Washington. The considerable distance from forests is reflected by the high proportions of grass pollen throughout.

sediments that are seasonally inundated. Thus, the paleic trends of grasses, Chenopods, and Composites become more significant when defined in terms of climate. The over-representation of forest trees in those sedimentary columns lying south and west of the forest boundary is well shown by an abundance of their pollen throughout the profiles. The uppermost horizons reveal as much as 25 per cent of yellow pine pollen, and yet these forests are 30 miles distant (figs. 72, 73). All of the sedimentary columns obtained from this area contain the volcanic ash stratum which permits chronological correlation of the several profiles and also correlation with those from other regions that contain the ash. Also, all sedimentary columns lie directly upon glacial drift or its chronological equivalent and are presumably about the same age.

LODGEPOLE PINE

Lodgepole pine was the principal arboreal invader of deglaciated terrain and also was apparently the most abundant tree persisting for some distance south of the glacier. Its pollen proportions at the bottom average 45 per cent, and range from 29 to 61 per cent (fig. 74). How near it grew to the sites of those sediments farthest south of the ice front is problematical. In the Crab Lake and Harrington profiles, the former

of which is located 30 miles from the glacial boundary, lodgepole is recorded to 54 and 61 per cent respectively at the bottom, and to 14 and 10 per cent at the top. At present the nearest forests containing lodgepole pine are at least 25 miles distant. This suggests that glacial and early postglacial forests grew much closer to the sites of these sediments than at present. The proportion of 61 per cent for lodgepole in the Harrington profile is almost as great as that at the bottom of many of the Puget Sound region profiles where it is assumed that lodgepole thrived immediately adjacent to the sites of the sediments. The factor of possible water transportation of pollen, however, must be considered in interpreting early postglacial forest movements upon the basis of the pollen profiles. During early postglacial time the streams were probably larger than at present, and were a more important factor in pollen transportation. This suggests greater over-representation of lodgepole pollen in the lower than in the upper horizons, and lesser southward expansion of lodgepole during glaciation than the pollen profiles might indicate.

The general over-all trend of lodgepole pine is one of gradual decline. There are many fluctuations in each profile, some of them correlative with one another and those of other species. The most rapid

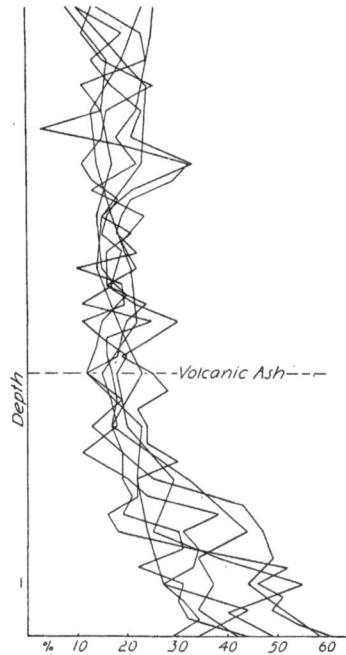

FIG. 74. Typical lodgepole pine pollen profiles of seven sedimentary columns located in the Columbia Basin, including four from the Channeled Scablands.

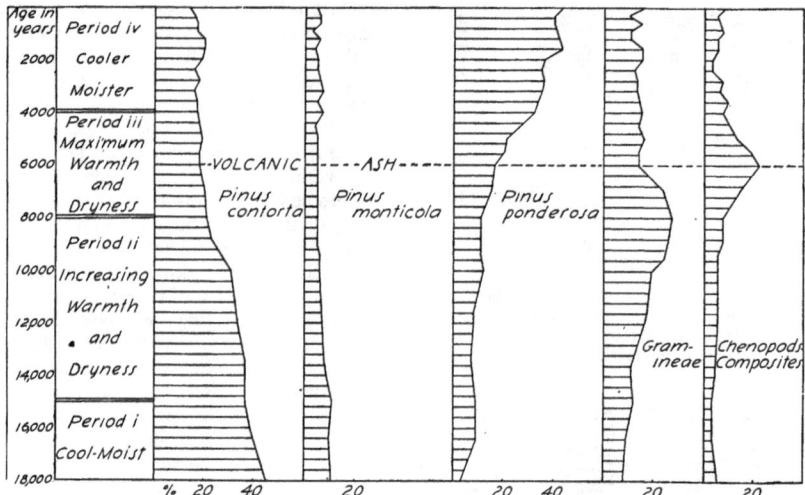

FIG. 75. The average of seven pollen profiles of Columbia Basin sedimentary columns. The sharp and probably meaningless fluctuations from level to level in individual profiles are eliminated and a better defined picture of postglacial vegetation is presented.

rate of decline in most profiles is during the time of warming and drying, and this is slightly reflected in the average pollen profile of the seven columns below the volcanic ash stratum (fig. 75). From this level upward the average pollen proportion shows a constant trend. Each profile shows many fluctuations from level to level, which must be ascribed to the peculiarities of the method. The average proportion at the uppermost level is 14 per cent, with a range from 8 to 25 per cent. In comparison with the Puget Sound region lodgepole was not as abundant during early postglacial time but it did not decrease so rapidly as to become almost non-existent. It apparently has not had to compete with such shade-tolerant species in eastern Washington as it did west of the Cascades owing to the drier climate during most or all of postglacial time. It still persists in local abundance within range of pollen dispersal to the scabland profiles, and immediately adjacent to those in the timbered Arid Transition.

WESTERN YELLOW PINE

During postglacial time the chief competitor of lodgepole has been yellow pine. In the Puget Sound region those species that replaced lodgepole as the glacial conditions were moderated were practically absent in the initial stages, as revealed by the lower strata. In eastern Washington, however, yellow pine which replaced lodgepole to a large extent since deglaciation was already abundant in the area when the pollen-bearing sediments were first deposited. It apparently was able to persist south of the glacier with

lodgepole, but because of rigorous climatic conditions, with perhaps too much moisture, it was not as prevalent. If the climate became progressively drier south of the glacier, yellow pine should have existed farther south than lodgepole because it is more xerophytic. The average of the seven profiles shows that yellow pine was recorded to 21 per cent at the lowest level, with a range of 0 to 33 per cent (fig. 76). In general the postglacial trend of yellow pine has been one of slow, gradual increase until the warm, dry period, when it remained constant for some time prior to the volcanic eruption. It then had a period of more rapid increase to a maximum which it has generally maintained to the present. Its average proportion for the uppermost level is 53 per cent with a range from 24 to 76 per cent. The two sedimentary columns located at some distance from the yellow pine forests contain a postglacial record for yellow pine that differs slightly from those within the forests. In the Harrington profile it increases with rather broad fluctuations to a maximum well above the ash stratum, and then declines to the surface (fig. 72). In the Crab Lake profile it remained more or less static, and in much lower proportions during the entire postglacial period, fluctuating between 12 and 29 per cent (fig. 73). These two profiles tend to distort the average trend of yellow pine in the timbered area from the time of ash deposition to the present. When they are not used to compute the average, a stronger trend for this species is revealed. The postglacial tendency of yellow pine suggests that it gradually replaced lodgepole pine as the climate became warmer and

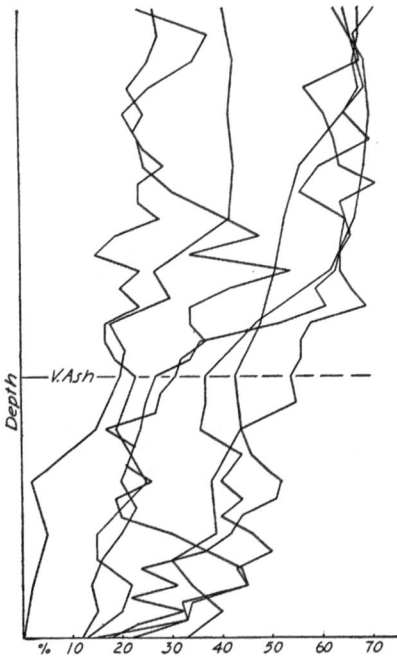

FIG. 76. Typical western yellow pine pollen profiles of seven sedimentary columns from the Columbia Basin, revealing the gradual expansion from early postglacial time to the present with the exception of those sections located in the timberless area. Most profiles show a retardation of yellow pine expansion during the warm, dry interval.

drier and the glacial environment was ameliorated. Its expansion was arrested by the warm, dry conditions from 8,000 to 4,000 years ago, and then it resumed its expansion to the maximum as the climate became moister and cooler, and maintained it to the present. Today it exists as the climatic climax dominant of the timbered Arid Transition area, a status which it has held for at least 4,000 years.

WESTERN WHITE PINE

It seems probable that much of the white pine pollen interred in the sedimentary columns has been transported from some distance both by the air and water. There is also the possibility that the size-range method of separating lodgepole and white pine pollen breaks down in certain areas. The range of white pine pollen proportions at the bottom is 5 to 17 per cent; it then fluctuates between 0 and 23 per cent to the top, where it ranges from 0 to 18 per cent (fig. 77). In general, white pine was more abundant and nearer the sites of the sediments in early postglacial time than later. This would be logical in view of the cooler and moister climate that must have existed. The profiles

reveal a general decline of white pine upward, with an accelerated gradient of decrease during the onset of the warm, dry interval, followed by a constant trend or in some instances a slight increase. An increase from the time of ash deposition almost to the top in the Newman Lake profile is correlative with its postvolcanic eruption increase in northern Idaho (fig. 78) and suggests the transportation of white pine pollen by water and wind from montane forests to the north and east.

GRASS

Grass pollen profiles of seven sedimentary columns from the sagebrush area and yellow pine forests dis-

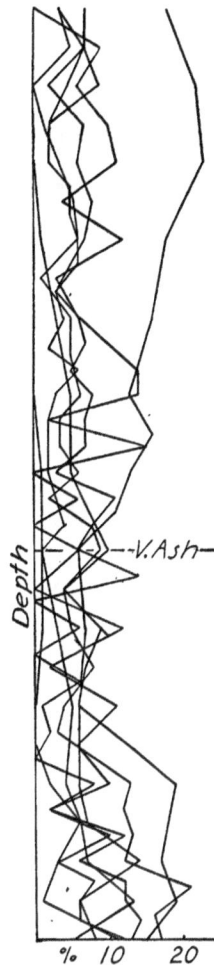

FIG. 77. Western white pine pollen profiles of seven sedimentary columns located in the Columbia Basin, showing generally greater abundance in the lower levels.

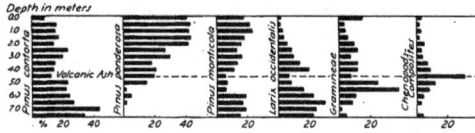

FIG. 78. Pollen diagram of the sedimentary column from Newman Lake near Spokane.

close excellent evidence for the development of the
postglacial warm, dry period. As already mentioned,
the grass pollen was not separated even as to genera,
and it probably represents grasses from both the
Upper Sonoran and timberless Arid Transition.
Those of the former life zone undoubtedly are more
xerophytic, and if their pollen could be distinguished
greater refinement in interpreting the pollen profiles
might be possible. However, it seems logical to
assume that an increase in grass pollen, especially in
those profiles located in the yellow pine forests, denotes drying and/or warming of the climate. The
development of the dry interval began soon after
initial sedimentation, and apparently reached its
maximum prior to the volcanic ash depositions. In
all except the Crab Lake profile grass attained its
maximum below the volcanic ash stratum (fig. 79).

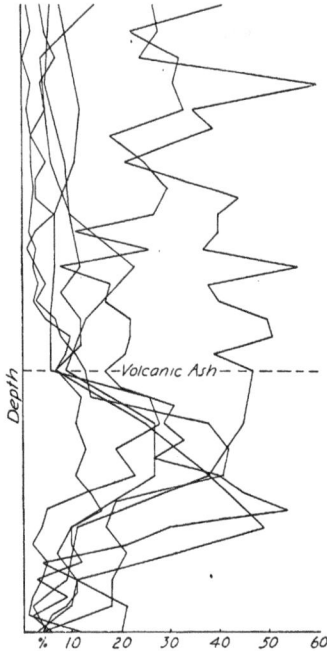

FIG. 79. Grass pollen profiles of seven sedimentary columns located in the Columbia Basin, revealing maximum attained
before the volcanic activity in all except one profile.

The maximum occurs at different depths in each
column, but owing to different rates of deposition it is
assumed that all maxima are approximately synchronous. The average for the lowest horizon is 8 per cent
with a range of 3 to 20 per cent. A gradual and almost constant increase is recorded, to an average maximum of 27 per cent, perhaps 2,000 years before the
volcanic activity responsible for the ash stratum.
The maximum in any one profile is 54 per cent in that
near Harrington (fig. 72). Grass more sharply
declined from its maximum than it had previously
increased to the time of ash deposition, and then remained almost constant to the present, with a maximum fluctuation of only 5 per cent. In only the
Crab Lake profile did grass maintain its predominance where it is recorded to 37 per cent at the uppermost horizon (fig. 73). Its maximum proportion of
60 per cent is attained at 1 meter. The other profile
located in the Upper Sonoran life zone, near Harrington, reveals that grass declined after its pre-volcanic
eruption maximum, and then again increased upward
in the profile to 41 per cent at the surface. As this
increase is not consistently recorded in the other profiles, it seems to be anomalous, as there is no evidence
in the yellow pine profiles to suggest that this species
had given way to a grass invasion. In general, grassland gradually expanded as the postglacial warming
and drying progressed, to reach a maximum at the
culmination of warmth and dryness. It then declined as the climate became cooler and moister, but
was partially replaced by Chenopods and Composites
before the peak of the warm, dry interval had passed.

CHENOPODS–COMPOSITES

Admittedly, these two groups of plants represent a
wide range of environment and thus their pollen record may offer more than one interpretation with respect to climatic trends. In general, it is assumed
that those species represented in the sedimentary
columns are much the same as exist today within
range of pollen dispersal, and their postglacial record
may be interpreted in terms of their present distribution and characteristics. Of the two groups, the
Chenopodiaceae are much more strongly represented.
Pollen of the Compositae family consists mostly of
that of the tribes Anthemideae and Ambrosieae, and
no attempt was made to separate the genera and the
species because of its scarcity. The greater abundance of Chenopod pollen requires an interpretation
based upon their recorded trends, even though as a
group they are fewer and more localized than the Composites. It is believed, however, that the Chenopod
trends also represent climatic changes indirectly reflected by an expansion of alkali areas due to increased
evaporation of standing water and drying up of lakes
and ground water.

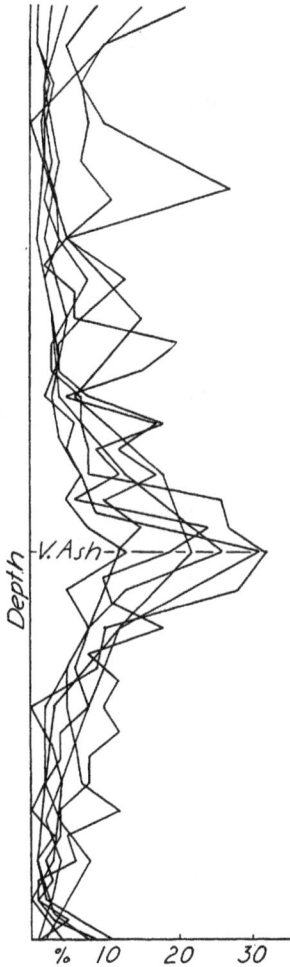

FIG. 80. Chenopod-Composite pollen profiles of seven sedimentary columns located in the Columbia Basin, showing the maximum proportion attained at the volcanic ash horizon in all except one profile.

The seven sedimentary columns of this group disclose a marked correlation of the Chenopod-Composite pollen profiles (fig. 80). The maximum of most profiles occurs at or near the volcanic ash stratum, and some time later than the peak of the grass pollen. At the bottom the average is 5 per cent with a range from 1 to 11 per cent. The average profile then shows very little fluctuation until the time of grass maximum, but as grass declined, Chenopods-Composites abruptly increased to their maximum. ‧ They attained an average of 22 per cent at the volcanic ash level (fig. 75). They declined as sharply after the volcanic

activity, and then remained more or less constant to the surface, fluctuating between only 7 and 9 per cent. At the surface the average pollen proportion is 8 per cent, with a range from 1 to 21 per cent. It is difficult to say whether the relative trends of grasses and Chenopods-Composites are competitive or complementary. The maximum of the latter, occurring later than that of grass, suggests persistence of the xeric period, but probably not increased xerism, to a maximum at the time of ash deposition. This view is supported by the resumption of yellow pine increase after a period of arrested expansion during the most rapid expansion of grass. If the maximum of Chenopods indicates increased xerism to its greatest degree at this time, then yellow pine should show either a static or a diminishing trend to this point, as it is not as xerophytic as the Chenopods and Composites. After Chenopods-Composites attained their greatest proportions, they declined to a lower proportion than did grass, and from that time on grass has been predominant (fig. 75). The relative degree of pollen preservation, amount of pollen produced, and what species are best represented, as well as other factors, are unknown and cannot be considered here, and thus the interpretations must be regarded as tentative. In general, it is believed that the relative indicated trends of the species concerned reflect warming and drying of the climate to a maximum perhaps 1,000 years before the ash was deposited, or about 7,000 years ago, which was maintained until approximately the time of the recorded volcanic activity. Then followed cooling with an increase in moisture, sufficient to permit yellow pine to resume its postglacial expansion that had been held in check during the xeric period.

OTHER SPECIES

The most abundantly and consistently represented of the other species or groups is *Abies*. All species of fir are considered collectively, as individually they are too sparsely and sporadically represented to hold any significance. In general fir is most strongly and constantly represented in the lower levels of all the profiles, suggesting their existence within range of pollen dispersal during the earlier and cooler period of postglacial time. Their subsequent diminution supports the other evidence for a xeric period. Perhaps the other species most significantly represented is western larch, in the Newman Lake profile (Hansen, 1939b). It shows two periods of influx, each followed by decline, suggesting severe recurring fire within range of pollen dispersal to the bog. One period of larch expansion occurred before the ash deposition, and the other period after the volcanic activity. The post-volcanic eruption influx may be synchronous with that recorded in a bog near Bonners Ferry, in northern Idaho (Hansen, 1943e). If such an increase was the result of fire, however, the fires must have been

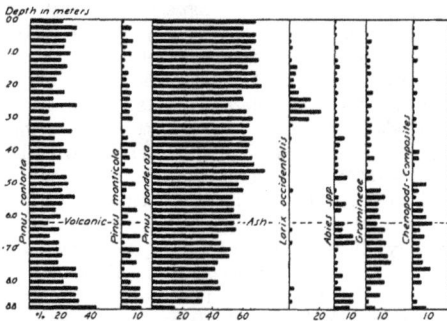

FIG. 81. Pollen diagram of the sedimentary column at Fish Lake near Cheney, Washington, on the edge of the Channeled Scablands, and just within the yellow pine forest zone. The recorded forest succession is typical of that within the timbered Arid Transition life area of Washington.

rather widespread or concurrent in local areas where the results could be reflected in each respective profile. An influx of larch to 21 per cent at 2.8 meters in the Cheney profile (fig. 81) may have been synchronous with a second period of expansion at Newman Lake and Bonners Ferry.

Other forest tree species represented sparsely and sporadically are Douglas fir, Engelmann spruce, western hemlock, and mountain hemlock. Douglas fir is most abundantly and consistently recorded, it being an associate of yellow pine in the upper part of the timbered Arid Transition area. None of these species is recorded to a significant degree. Alder, willow, maple, birch, and cottonwood are the broadleaf species whose pollen is contained, and no significance with respect to climate or forest succession can be attached to their trends. Hydrarch succession at the site of the sediments is portrayed by the pollen record of hydrophytic species. (See figs. 82, 83, 84 for additional eastern Washington profiles.)

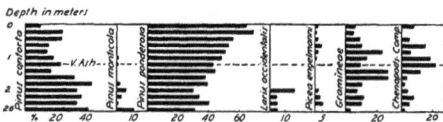

FIG. 82. Pollen diagram of the sedimentary column near Wilbur, Washington, in the Channeled Scablands.

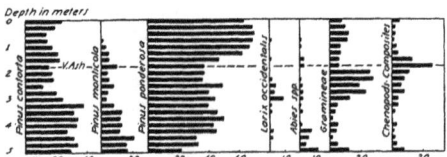

FIG. 83. Pollen diagram of the sedimentary column at Eloika Lake in eastern Washington in the yellow pine zone.

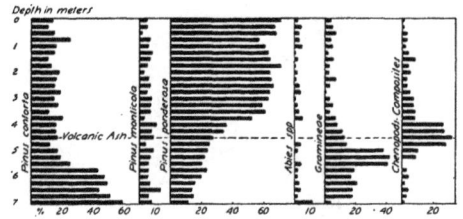

FIG. 84. Pollen diagram of the sedimentary column at Liberty Lake, near Spokane, Washington.

POSTGLACIAL FOREST SUCCESSION IN NORTHERN WASHINGTON AND IDAHO

The five sedimentary columns comprising this group are not all located within the same phytogeographic province. Two of them from near Priest Lake and Bonners Ferry in northern Idaho are situated in the subclimax western white pine forests. The other three, two of which occur on the east slope of the Cascades and the other in the Okanogan Highlands, are situated near the upper limits of the yellow pine forests. All profiles contain the volcanic ash stratum, and can be fairly well correlated as to the major trends of the species represented.

LODGEPOLE PINE

In the three northernmost profiles lodgepole pine followed about the same general trends as in those of the Scablands and Arid Transition area of eastern Washington, discussed immediately above. It is recorded to its greatest abundance in the lower levels and gradually declines with sharp and broad fluctuations to the present (figs. 85, 86, 87). In the Idaho profiles its trends are contrary to those of western white pine and western larch which have been its chief competitors during postglacial time. It was apparently more abundant than white pine for more than half of postglacial time, but white pine gained supremacy after the warm, dry stage. In the vicinity of Bonaparte Lake lodgepole gradually gave way to yellow pine and Douglas fir, while near Wenatchee and Lake Wenatchee it was never abundant and

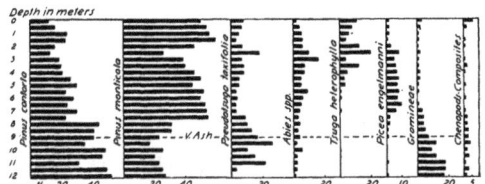

FIG. 85. Pollen diagram of the sedimentary column near Priest Lake in northern Idaho, in the western white pine forests. The predominance of lodgepole and white pine is shown throughout, with a brief influx of the climax dominants in the upper third of the profile.

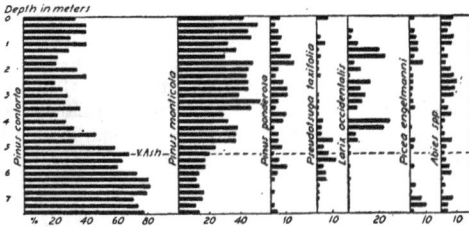

FIG. 86. Pollen diagram of the sedimentary column near Bonners Ferry in northern Idaho. Lodgepole and western white pine predominance are indicated throughout, with two significant expansions of western larch occurring above the volcanic ash stratum.

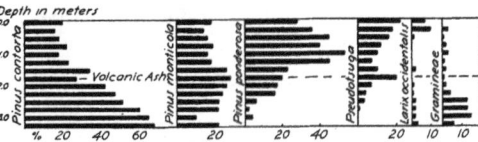

FIG. 87. Pollen diagram of the sedimentary column at Bonaparte Lake in the Okanogan Highlands of northern Washington. The appreciable proportions of grass pollen in the lower levels may reflect the influence of the warm, dry stage, more strongly manifested in the Columbia Basin.

persisted in about the same proportions during the entire time represented by the sedimentary columns (figs. 88, 89). The absence of recorded lodgepole predominance in the lower levels of these profiles may signify the persistence of yellow pine on adjacent unglaciated ridges during late-glacial time, which is also reflected by maximum grass pollen proportions in the lower levels. In these three profiles the proportions of white, lodgepole, and yellow pine, and Douglas fir are made relatively lower by the presence of fir, hemlock, and spruce pollen which probably drifted down from the Canadian and Hudsonian life zones, located on the adjacent steep mountain slopes.

WESTERN YELLOW PINE

In the two northern Idaho profiles yellow pine has played a minor role in postglacial forest succession. At Bonners Ferry its proportions are low throughout, attaining their maximum above the volcanic ash stratum and after the xeric interval. At Priest Lake yellow pine was largely absent in the adjacent forests, as its highest proportion is only 3 per cent. This area has sufficient precipitation to support more mesophytic species, and yellow pine being very intolerant of shade has evidently been unable to compete under the existent climate. Yellow pine is not abundant in the present forests.

In the other three profiles, from bogs located within the yellow pine forest, yellow pine has followed essentially the same trends as in the Scabland and Transition

areas at lower elevations farther south and east. In the Bonaparte Lake profile yellow pine declines from its maximum above the volcanic ash stratum, which is the converse of Douglas fir increase, denoting moister and cooler conditions in more recent time. Near Lake Wenatchee the abundance of yellow pine in the lower horizons suggests its occupancy during glaciation of adjacent unglaciated ridges where it still persisted during early postglacial time. In general, the postglacial trends of yellow pine in these montane profiles indicate its gradual replacement of lodgepole and its expansion to a predominant status in a climatic yellow pine climax which was held in check during the xeric period.

WESTERN WHITE PINE AND WESTERN LARCH

The postglacial successional trends of western white pine in montane areas of northern Idaho and Washington are apparently an expression of its successional relationships to its associates, in relation to the effects of forest fires. As previously stated, white pine has persisted as a subclimax species in a forest where western hemlock, western red cedar, and lowland white fir are the climax dominants due to recurring fire. Its successional relationship to the subdominants western larch and lodgepole pine is also greatly influenced by fire, as these two species precede white

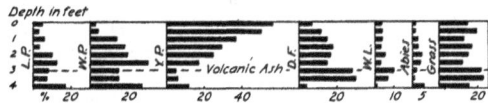

FIG. 88. Pollen diagram of the sedimentary column near Wenatchee, Washington, on the eastern slope of the Cascade Range. The influence of the grasslands immediately adjacent to the east is shown by the abundance of grass pollen throughout. L.P., lodgepole pine; Y.P., yellow pine; D.F., Douglas fir; W.L., larch.

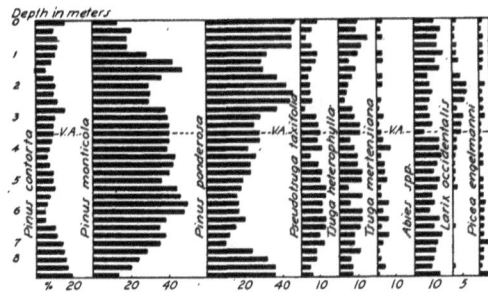

FIG. 89. Pollen diagram of the sedimentary column at Fish Lake, near Lake Wenatchee, on the east slope of the Cascades in the yellow pine forests. The influence of several life zones within range of pollen dispersal to the site of the sediments is shown by the large number of montane forest trees represented in the profile.

pine after severe fires and are eventually eliminated from the forest complex in the continued absence of fire. Douglas fir, of questionable status, seems to be more closely associated with the climax than other members of this forest complex. The postglacial fluctuations of white pine have been largely opposed to those of lodgepole and larch and partially controlled by climate. In the forests about Priest Lake and Bonners Ferry white pine was abundant in early postglacial time, but not predominant, and is recorded to 27 and 13 per cent respectively. Its steady postglacial expansion was arrested during the warm, dry interval until the time of volcanic activity. It then increased rapidly, superseded lodgepole soon after the volcanic eruption, and remained almost continuously predominant to the present. During this period of white pine predominance its supremacy was threatened by influxes of both lodgepole and larch, presumably as a result of the effects of severe and recurring fires. This is best shown in the Bonners Ferry profile where larch benefited at least twice as a result of fire (fig. 86). The first period was prior to the recorded volcanic activity, when larch attained a maximum of 30 per cent, to exceed white pine. During the second period of larch expansion it reached a proportion of 24 per cent at 1.5 meters. The significance of these maxima attained by larch is much magnified when it is considered that the amount of pollen produced by this species is comparatively small. Soon after the second period of larch influx fire of lesser scope and intensity favored a temporary increase of lodgepole to supersede white pine at one level.

Although western hemlock, western red cedar, and lowland white fir are the dominants of the climax forest, they have been unable to attain predominance in northern Idaho during postglacial time. During the early part the rigorous environment permitted lodgepole pine to thrive. Then the development of a warmer and drier climate hindered their expansion, probably with the aid of fire. Since the dry period, fire has apparently favored the persistence of lodgepole, white pine, and larch. In the Priest Lake profile hemlock, Douglas fir, and lowland white fir expanded during the time represented at depths of 3 to 2 meters in the sedimentary column, but this expansion was short-lived (fig. 85). A recent and present-day advance of hemlock and Douglas fir is recorded in the upper levels. At Bonners Ferry, the pollen record does not reveal a climax trend at any point in the profile. In general, the postglacial trend of white pine was one of slow initial development due to a rigorous environment; then its expansion was retarded during the dry period, and followed by a period of predominance to the present encouraged by increased moisture and periodic pyric influence.

In the other three sedimentary columns, located in the yellow pine forests, the recorded trends of white pine cannot be interpreted in the same way as those in northern Idaho. Near Lake Wenatchee it has been predominant during most of postglacial time, even during the dry period (fig. 89). Its principal fluctuations have been the opposite to those of yellow pine, and theoretically should reflect both climatic trends and the influence of fire. The pollen identified as that of white pine may include considerable whitebark pine pollen, and the pollen of both may have drifted down from the Canadian and Hudsonian life zones. In the Bonaparte Lake profile white pine is recorded constantly throughout the profile, while near Wenatchee a short interval of predominance after the time of volcanic activity suggests fire. Pollen from higher life zones is probably interred in the sedimentary columns, somewhat distorting the true picture of adjacent forest succession.

OTHER CONIFERS

Other coniferous species recorded to a limited extent include fir, Engelmann spruce, western red cedar, and mountain hemlock. Of these, two or more species of fir are most consistently and abundantly recorded. In general, their highest proportions occur above the ash stratum, suggesting that the rigorous early postglacial environment and the succeeding dry period were unfavorable. Engelmann spruce is also most abundantly recorded in the upper levels of the profiles, suggesting a downward movement of the Canadian and Hudsonian life zones after the dry period had passed. With the exception of lowland white fir, the pollen of these species undoubtedly came from higher elevations during most of postglacial time. The maximum of lowland white fir was attained at Priest Lake during the abortive climax development in recent time.

GRASS

In all of the profiles except that at Bonners Ferry grass is represented in significant proportions below the volcanic ash layer. In the Priest Lake profile it is recorded from 9 to 18 per cent up to the ash level and then declines to low proportions throughout the remainder of the profile. In the Bonaparte Lake profile grass is also represented to its maximum proportion of 15 per cent below the ash, while the same is true for the Wenatchee profile in which its maximum is 23 per cent. These respective maximum proportions of grass, recorded in profiles located in forested areas, approximately synchronous with those of the Scabland and Transition areas to the south and east are of extreme significance in corroborating the evidence of the warm, dry climate of middle postglacial time. Although Bonaparte Lake is 50 miles or more distant from the present prairie region of east central Washington, the influence of the dry period was recorded and adds considerable support to the evidence for its existence manifested by other Pacific Northwest pollen profiles, as well as to evidence from other sources. Chenopods and Composites, while recorded to low proportions, also apparently were most abundant before the time of volcanic activity. The grass maxima are recorded at relatively lower stratigraphic positions than in columns of the Scablands and Transition area, suggesting that the former are slightly younger. This would be consistent with the northerly retreat of the glacier, permitting earlier sedimentation near the southern limits of glaciation as well as in the unglaciated region. The other alternative would imply a slower rate of sedimentation.

FOREST SUCCESSION ON THE PACIFIC COAST

GENERAL CONSIDERATIONS

The factors that have controlled forest succession on the Pacific Coast of Oregon and Washington for that part of postglacial time represented by the sedimentary columns are not so clearly depicted as elsewhere in the Pacific Northwest. The pollen profiles of a single species from several bogs do not show such positive correlation as those from other regions, while the recorded trends of forest succession do not seem to agree chronologically in all areas. This may be due to the different ages of the bogs, the varying forest composition and stages of succession during initial sedimentation, differences in topography, varying degrees and periods of sand movement, and the position of bogs in relation to sand dune areas, the ocean, the Coast Range, direction of prevailing winds, and the adjacent forests. The absolute and relative ages of the coast pollen-bearing profiles are indefinite, as they lie beyond the limits of Pleistocene glaciation. One or two of the profiles may have originated during late-glacial or early postglacial time, but the ages of these are less certain than those of bogs that rest directly upon glacial drift. The other sedimentary columns, because of the physiographic method of formation of their basins, may vary considerably in age. There is no time marker present such as the volcanic ash stratum in Washington profiles, so there is no point at which to begin stratigraphic correlation of the several profiles except from the top. Correlation begun at the top and continued downward level for level discloses no evidence for a systematic response of the several species to environmental changes over the entire area. The disparity in the rate of organic sedimentation may tend to obscure what positive correlation does exist. In certain groups of profiles, a correlation does exist although the profiles involved may not be from the same geographic area.

There is no reason to assume that the coastal strip was not forested during the Pleistocene, and if the sea level was lower during this time a wider zone of coastal forests may have existed. These forests were probably of the climax type, and were the result of some thousands of years of reaction upon the environment.

Locally, the forest cover may periodically have been in a stage of flux due to physiographic and edaphic instability. Possibly at the time of origin of certain sedimentary columns the adjacent forests were stabilized and composed largely of the climax species, while at other sites the forests within range of pollen dispersal were undergoing succession due to recent sand dune movement or even to fire. This is suggested by the pollen proportions in the lowest levels of the several sedimentary columns.

Another factor which has prohibited regional changes in vegetation that would be represented systematically in the sedimentary columns is the relative constancy of postglacial climate along the coast. The incidence of the warm, dry period as indicated by postglacial vegetative succession in Washington also serves as a chronological marker, but this source of chronology is absent in the coast bogs. The postglacial climate of the north Pacific Coast has probably been essentially marine, with little variation, even during the dry period that developed farther inland. While climatic fluctuations undoubtedly did occur, they were of insufficient magnitude to cause a systematic response by the vegetation. Apparently the precipitation did not drop below nor the temperatures rise above the optimum for the dominant species, spruce and hemlock, and they were not replaced by xerophytic species. There is, of course, the possibility that the warm, dry period occurred before most of the profiles were initiated if we estimate their ages at less than 6,000 years. Those profiles, however, that may represent most or all of postglacial time do not seem to record the dry period

Perhaps the principal causes of forest succession and retrogression have been the periods of physiographic and edaphic instability caused by shifting sand. This is evident today and undoubtedly has often occurred in the past. It is not possible to say whether the prehistoric periods of sand movement were local or general for the entire coastal strip. It may be that some of these periods of physiographic instability were regional and were a result of climatic changes that are more definitely expressed in profiles from farther inland. Periodic movement of sand buried forests in various stages of succession initiated on sand dune as well as on mature bogs. The pioneer forests of lodgepole were probably more liable to be destroyed because of their proximity to the ocean and their establishment on less stabilized soil than the climax forest. The destruction of the lodgepole forests is perhaps more strongly reflected in the pollen profiles than that of the climax forests of spruce and fir because the bogs usually lie leeward to the pine zone. Thus, the fluctuations of the three species in the pollen profiles, when opposed to one another or one group to another, may denote relative rather than absolute changes in abundance. It seems probable that fluctuations in the pollen profiles are more often a result of changes in the abundance of lodgepole rather than in spruce or hemlock. The location of the pine forests, largely windward to the bogs, has likely accentuated their degree of over-representation caused by pine's greater pollen production. Spruce and hemlock forests located largely leeward to the bogs are generally under-represented in the pollen profiles.

An additional factor that has influenced materially the pollen proportions, and in some cases tended to distort them, is the arboreal invasion of the mature bog. In some areas along the coast there are hundreds of acres of shallow mature bogs, supporting various stages of arboreal succession. The development in the past of forests on mature bogs, long since buried or obliterated by climax forests, is revealed by exhumation of peat strata overlain by many feet of sand (Hansen and Allison, 1942). The three principal species of the forest, lodgepole pine, spruce, and hemlock, all invade the mature bog, and usually in that sequence and order of abundance. However, any one may be the pioneer invader and the most abundant. In the uppermost levels of profiles from bogs which have been invaded recently by lodgepole this species may be recorded to over 50 per cent although the adjacent forests are composed essentially of spruce and hemlock. In one instance the same is true for spruce. Although this process and its pollen record are evident for the present time and can be interpreted correctly, it is hardly possible to determine to what extent the pollen profile fluctuations representative of paleic forests may reflect this process. The following interpretation of the Pacific Coast pollen profiles is tempered by these considerations.

LODGEPOLE PINE

The pollen profiles of lodgepole pine differ in their general trends in most of the sedimentary columns, and they reveal no consistent trends such as those of the Puget Sound region or eastern Washington. In these regions conditions for its preponderance were favorable only during the later stages of deglaciation or early postglacial time, and then it was replaced by the dominants of the climax or subclimax forests. On the coast conditions have been favorable for at least its local predominance periodically during all the time represented by the sedimentary columns. The pollen profiles reveal five different kinds of trends for lodgepole. First, in some sedimentary columns it is recorded as never having been predominant and its pollen proportions are low and remain constant throughout the profiles. This is true of the profiles from Woahink Lake and at Grayland. (fig. 90). Second, in some profiles lodgepole pollen is not abundant below the upper levels where it may increase abruptly to 60 per cent or more. Its proportion in adjacent forests has been about the same as the first, but in recent time it has invaded the bog from which the

profile was obtained. Such has occurred on bogs near Gearhart and Sandlake, Oregon (fig. 91). A third trend of lodgepole is revealed by the profiles from near Hauser, Oregon, and Hoquiam, Washington (fig. 92). where lodgepole has followed the general

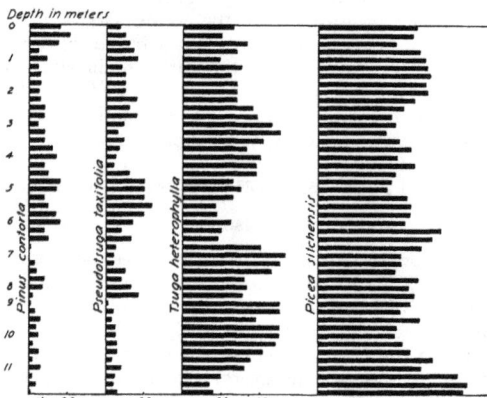

FIG. 90. Pollen diagram of the sedimentary column at Woahink Lake, on the Oregon Coast. The predominance of Sitka spruce throughout the time represented may be due to the location of the lake inland from the sand dune area immediately adjacent to the ocean. This section probably represents most of postglacial time.

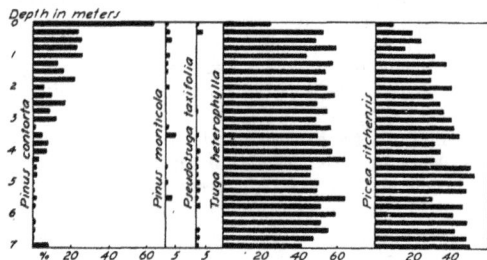

FIG. 91. Pollen diagram of the sedimentary column near Gearhart, Oregon, a few miles inland from the ocean. Although Sitka spruce and western hemlock have been predominant in adjacent forests, the recent invasion of lodgepole on the mature bog is reflected in the upper levels.

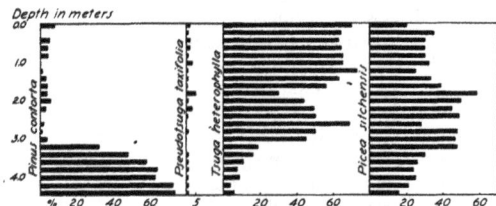

FIG. 92. Pollen diagram of the sedimentary column near Hoquiam, Washington, near the Pacific Ocean. The trend of forest succession is somewhat similar to that in the Puget Sound region, probably because of the influence of glacial outwash from the Olympic Mountain glaciers.

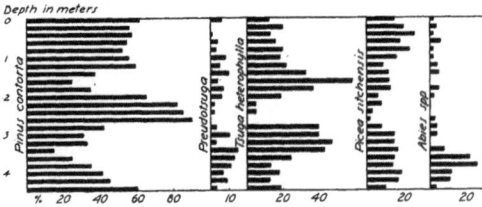

FIG. 93. Pollen diagram of the sedimentary column near Bandon on the southern Oregon coast. The influence of the extensive dune areas in the vicinity is shown by the predominance of lodgepole most of the time. Periods of extensive sand movement are suggested by the wide fluctuations of lodgepole pine.

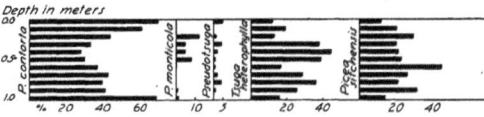

FIG. 94. Pollen diagram of the sedimentary column near Newport on the Oregon coast. The presence of early lodgepole pine forests on adjacent dunes, and the invasion of the mature bog by lodgepole in recent time are well shown.

trend portrayed in Puget Sound profiles. In the lower levels it is recorded to 60 per cent or more, and then sharply decreases to 20 per cent or less, and is recorded to low proportions throughout the rest of the sedimentary column. This type of lodgepole pollen profile suggests favorable · conditions when sedimentation began, perhaps owing to sand movement. A fourth type of lodgepole pollen profile is revealed by sedimentary columns near Bandon and Marshfield, Oregon (fig. 93). Lodgepole is recorded to proportions over 60 per cent at the bottom, and then fluctuates widely, reaching 90 per cent in the middle of the Bandon profile, and maintaining high proportions throughout. Its broad fluctuations suggest periodic sand movement, destroying the lodgepole pine forests, which is reflected by its low proportions, or they may reflect an invasion of mature bogs that existed at those times. A fifth trend of lodgepole is depicted by the profile from near Newport, Oregon (fig. 94), where it is recorded to 69 per cent at the bottom, declines to a low of 28 per cent in the middle, and then sharply expands to 66 per cent at the surface. Apparently lodgepole was abundant in adjacent sand dune areas when sedimentation began, then was gradually replaced by the climax forest species, and finally as the bog matured it invaded the surface and is probably over-represented as far as the adjacent forests are concerned.

SPRUCE AND HEMLOCK

Observations of present-day coastal forests indicate that Sitka spruce succeeds lodgepole pine and in turn is succeeded by western hemlock on dunes and other sandy areas. Spruce also precedes hemlock on the

mature bog, indicating that it can thrive under less favorable conditions than hemlock. In general, spruce and hemlock have been more abundant than lodgepole, except for brief periods when apparently local conditions were favorable for an influx of lodgepole. Throughout the profiles the trends of spruce and hemlock have been consistently opposed to each other, indicating that these two species have been directly competitive rather than complementary in their successional trends. Lodgepole and spruce have been more complementary to each other, while only in a few cases have lodgepole and hemlock succession been similar. This indicates that lodgepole and spruce preceded hemlock in forestation of new areas formed by shifting sand, and were often associated in the early stages. Spruce being more tolerant and longer-lived, however, is able to persist while lodgepole is finally eliminated if succession is not interrupted. In a few columns spruce and hemlock trends are similar and opposed to those of lodgepole. In general, spruce attains its higher proportions in the lower half of the profiles, and hemlock increases to supersede spruce and maintain predominance in the upper part

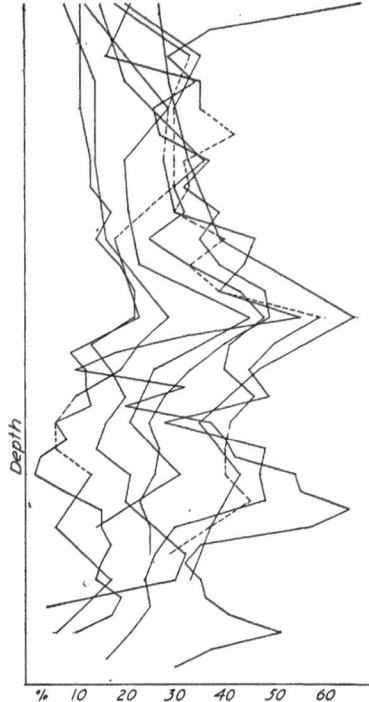

FIG. 95. Sitka spruce pollen profiles of nine sedimentary columns on the Pacific Coast, showing an expansion of this species which may have been approximately synchronous in all profiles. Other expansions and declines are not correlated because of possible differences in rates of sedimentation.

of the profiles. This suggests that modification of the environment, particularly the soil, has favored the expansion of hemlock. In attempting to show some logical correlation it was found that spruce is recorded in nine of the eleven profiles as having attained its maximum at levels from one-half to two-thirds upward in the columns (fig. 95). In correlating the profiles of lodgepole and hemlock for the same levels above and below this maximum, however, little correlation was found to exist. In the Bandon and Marshfield profiles, lodgepole pine shows somewhat similar trends (fig. 96). Those profiles where spruce and lodgepole are predominant at the top reflect the invasion of the bog and constitute an over-representation of these species as far as the composition of the adjacent forests is concerned. In generalizing the interpretation of the trends of lodgepole, spruce, and hemlock the periodic predominance of lodgepole, whether at the time that sedimentation began or later, signifies the accelerated movement of sand which destroyed the climax forest and permitted the influx of lodgepole to maintain temporary and local predominance. Possibly the same factor that fostered the expansion of lodgepole was instrumental in destroying it before the climax forest had begun to replace it. Thus, accelerated movement of sand may have buried lodgepole pine forests and caused a rela-

tive increase in the pollen of spruce and hemlock. A new period of lodgepole invasion then paved the way for the climax forest, first spruce, followed by hemlock. The periods of spruce predominance, usually succeeding those of lodgepole, suggest the elimination of lodgepole by the more tolerant and longer-lived spruce. When the physiographic and edaphic conditions remained static, hemlock invaded and remained predominant until a new period of sand movement began when it was temporarily replaced. Some of these periods of pine and spruce predominance may denote their invasion of adjacent bogs, since buried or obliterated by climax forest. In general, the environment has been favorable for hemlock predominance since the warm, dry interval, if it is assumed that most of the sedimentary columns are 6,000 years or less in age. Periods of accelerated sand movement, as reflected by the pollen profiles may have been synchronous with climatic fluctuations farther inland, but do not seem to be systematically indicated by the recorded trends of forest succession.

OTHER SPECIES

The principal other species represented by its pollen in the coast sedimentary columns is Douglas fir. It never has existed in abundance within range of pollen dispersal to the site of the sediments, and in no profile at any level does it assume predominance. In general, it seems to have been similar to hemlock in its successional trends, suggesting that it was able to thrive best during periods favorable for hemlock. It is most abundantly and consistently represented in the Bandon, Marshfield, Hauser, and Woahink Lake profiles. Its highest proportion is 25 per cent in the Woahink Lake profile.

The genus *Abies* is sparsely and sporadically represented and only in the Bandon and Marshfield profiles is it appreciably recorded (fig. 93). Its highest proportion of 26 per cent occurs in the Bandon profile, and this maximum is concurrent with pine and spruce predominance and immediately prior to a sharp increase in hemlock. No significance can be attached to the recorded trends of fir except that its general absence and low proportions throughout indicate that it was unable to compete successfully with the other species under the existent environment.

In the Forks profile on the Olympic Peninsula mountain hemlock is recorded in significant proportions. It reached over 20 per cent at the lowest level, slightly declined upward, and then sharply increased to 54 per cent in the middle of the profile. A marked increase of lodgepole immediately above and the occurrence of much silt at these levels in the sedimentary column suggest increased erosion. The conditions responsible for increased erosion may also have caused transportation of mountain hemlock pollen from a higher altitude in the Olympic Mountains to the east.

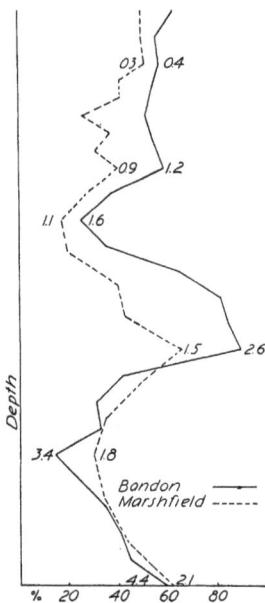

FIG. 96. Possible correlation of lodgepole pine fluctuations recorded in sedimentary columns near Bandon and Marshfield on the southern Oregon coast. Although the columns are of different thicknesses, it is possible that a depositional rate differential is involved.

The principal broadleaf species recorded is red alder. Its fluctuations seem to show no correlation with the trends of the forest tree species. In some instances its trend is similar to that of spruce and in others it follows that of hemlock.

POSTGLACIAL FOREST SUCCESSION IN THE CASCADE RANGE OF OREGON

Eight of the bogs located in the Oregon Cascades from which sedimentary columns have been obtained lie within the Canadian life zone, while one is near the upper limits of the timbered Arid Transition area of yellow pine belt and the other is in the Hudsonian life zone. Ranging in altitude from 2,200 to 6,200 feet, the recorded forest succession for the extremes varies somewhat, but, as most of the profiles are located between 3,000 and 5,000 feet, much the same general trends for the several species are to be noted. The difference in elevation is offset by the presence of the pumice mantle north and east of Crater Lake for a distance of more than 100 miles. The edaphic conditions afforded by the pumice have been favorable for predominance of lodgepole in most of the life zones, particularly the Canadian, and has tended to eliminate competition by more tolerant and longer-lived species. Five of the ten profiles rest directly upon Mount Mazama pumice and therefore record the forest succession since that time, perhaps for a period of not over 10,000 years, and possibly less, owing to the lapse of time between the pumice deposition and the initial sedimentation. Two of the profiles, those near Bend and on the Willamette Pass, were initiated before the eruption of Mount Mazama, and this volcanic activity is recorded by pumice strata in the sedimentary columns. The two remaining profiles, from Clear Lake and Clackamas Lake, are located north of the main pumice falls from Mount Mazama, Newberry Crater, South Sister, and Devil's Hill (map 2). Underlying volcanic glass, however, indicates a source of pumice within range by either water or air transportation during the early stages of sedimentation. These profiles, as previously mentioned, are situated within the boundaries of Pleistocene mountain glaciation, and are probably not much older than those that rest directly upon pumice of known age.

LODGEPOLE PINE

The influence of the pumice mantle on post-Mount Mazama forest succession is well shown in those sedimentary columns that rest upon this volcanic material. Lodgepole pine has been the predominant species in a region where the climate is favorable for a yellow pine climax in the Arid Transition area and a spruce-fir climax in the Canadian and Hudsonian life zones. This is effectively borne out by a comparison of sedimentary columns resting on the thicker pumice mantle with those that lie upon the thinner layers, and still

further with those lying outside the pumice-covered region. It is also well demonstrated by the change in the forest composition immediately above the pumice strata in those profiles that antedate the Mount Mazama eruption. In the areas about Munson Valley, on the south slope of Crater Lake Mountain; Big Marsh, 30 miles north of Crater Lake; and Diamond Lake, 12 miles north of Crater Lake; lodgepole pine has been strongly preponderant since the eruption of Mount Mazama, averaging close to 60 per cent throughout the pollen profiles (figs. 97, 98, 99). In the profile on Williamette Pass, west of the divide which was initiated before the volcanic activity, lodge-

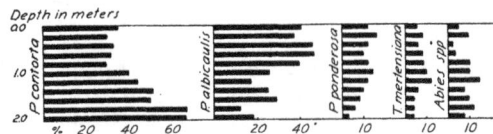

FIG. 97. Pollen diagram of the sedimentary column in Munson Valley, located a few miles below the south rim of Crater Lake caldera. Although the bog has an elevation of more than 6,000 feet, the influence of the pumice mantle is reflected in the predominance of lodgepole until the cooler and moister climate of the last 4,000 years.

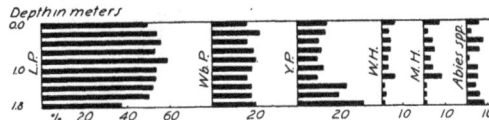

FIG. 98. Pollen diagram of the sedimentary column at Big Marsh, 30 miles north of Crater Lake, showing the influence of the pumic mantle in supporting lodgepole pine predominance throughout. L.P., lodgepole; Wb.P., whitebark pine; Y.P., yellow pine; W.H., western hemlock; M.H., mountain hemlock.

FIG. 99. Pollen diagram of the sedimentary column at Diamond Lake, northwest of Crater Lake, showing lodgepole pine predominance due to the sterile pumice. L.P., lodgepole pine; W.P., whitebark pine; Y.P., yellow pine; D.F., Douglas fir; W.H., western hemlock; M.H., mountain hemlock.

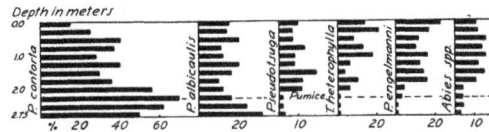

FIG. 100. Pollen diagram of the sedimentary column on Willamette Pass, Oregon. The influence of the pumice fall is reflected in the expansion of lodgepole from the bottom. The effect of the thinner pumice mantle and the moister climate of the west slope of the Cascades is apparent in the appreciable proportions of Douglas fir, western hemlock, and Engelmann spruce.

pole declined from the lowest level but immediately after the eruption it increased to 70 per cent as a result of the influence of the pumice fall. It then declined as the edaphic conditions, altered by only a thin layer of pumice, were rapidly modified and became suitable· for the climax species (fig. 100). In the Bend profile the postglacial trend toward a yellow pine climax after lodgepole had been predominant in early postglacial time was interrupted by the pumice fall, and lodgepole again expanded to predominance and remained so to the present (fig. 101). A second trend toward a yellow pine climax was interrupted by another pumice fall from a nearby volcano, and lodgepole pine again benefited correspondingly. In the Rogue River profile, which is located southwest of Crater·Lake, where the pumice is confined to the valley floors, lodgepole was never predominant during the time represented (fig. 102). The general predominance of yellow and sugar pine throughout indicates

Fig. 101.   Pollen diagram of the sedimentary column at Tumalo Lake, 13 miles west of Bend, Oregon. The influence of the Crater Lake pumice, occurring at 4.5 meters is shown by an expansion of lodgepole pine, after yellow pine, the climatic climax dominant, had fairly well replaced it. A second interruption of yellow pine climax trend is suggested at 2.0 meters where a stratum of pumice probably from Devil's Hill occurs.

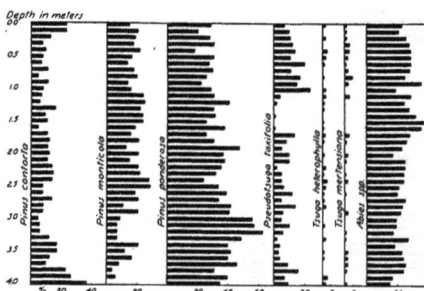

Fig. 102.   Pollen diagram of the sedimentary column in the upper Rogue River Valley, southwest of Crater Lake. Pumice in this region occurs chiefly on the valley floors owing to pumice flows, and the resultant low proportions of lodgepole pine in the upland forests is recorded. The western yellow pine profile may include negligible proportions of sugar pine and Jeffrey pine. The pollen of fir consists largely of white fir although no attempt has been made to segregate the several species of fir possibly represented.

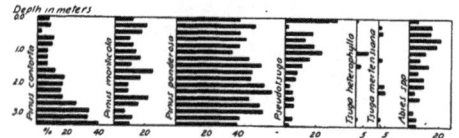

Fig. 103.   Pollen diagram of the sedimentary column near Prospect, in the upper Rogue River Valley, about 25 miles below the Rogue River profile. The general predominance of western yellow and sugar pine results from the confinement of the pumice to the valley floor.

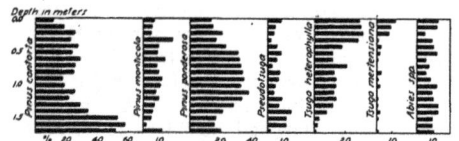

Fig. 104.   Pollen diagram of the sedimentary column at Clackamas Lake, in the Oregon Cascades. This area was glaciated but not covered with pumice, and the type of forest succession has been similar to that in other glaciated areas in the Pacific Northwest. The influence of greater moisture during the last 4,000 years is reflected by western hemlock expansion and yellow pine decline in the upper levels.

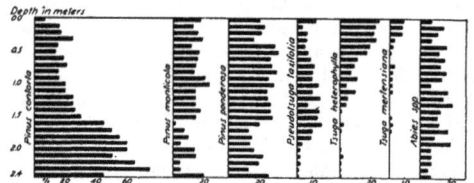

Fig. 105.   Pollen diagram of the sedimentary column at Clear ·Lake, located about 6 miles north of Clackamas Lake. The recorded forest succession in the two areas is similar.

that the edaphic conditions were not sufficiently altered to favor an expansion of lodgepole. The same trend is recorded by a profile from near Prospect, located in the same general area (fig. 103). At Clackamas and Clear Lakes the postglacial trend of lodgepole has been similar to that in other glaciated regions in the Pacific Northwest, as the influence of the pumice deposition apparently was not felt. It was predominant in the lower levels, representing early postglacial time, and then as postglacial environment was ameliorated other species of the climax forest partially replaced lodgepole (figs. 104, 105). In general, the postglacial forest succession in the Cascade Range of Oregon probably would have followed the same trends as in other regions of the Pacific Northwest if it had not been for the influence of volcanic activity over such a wide area, i. e., initial lodgepole pine predominance supplanted by the climatic climax dominants. Here it was interrupted, however, by the deposition of the pumice mantle and the resultant sterile soil.

WESTERN YELLOW PINE

The influence of the pumice is reflected in the profiles of yellow pine, but by trends opposite from those of lodgepole. In the areas of the thickest pumice mantle yellow pine has been far from predominant during the time represented. In the Willamette Pass profile it is only sparsely recorded because of the bog's location on the moister west slope of the Cascade Range. In the Bend profile yellow pine expanded to predominance in early postglacial time, and then was superseded by lodgepole as a result of the pumice deposition. A second climax trend was also interrupted by the influence of more recent volcanic activity. The best representation of yellow pine is in the Rogue River and Prospect profiles. In these regions it has been essentially preponderant during the entire time represented, reflecting the absence of the pumice mantle influence on adjacent forests. Its expansion from the bottom to its maximum a few levels above suggests response to the warm, dry stage which apparently followed the Mount Mazama volcanic eruptions. In the Clear Lake region yellow pine expanded in early postglacial time to supersede lodgepole as glacial conditions ameliorated. It was slightly replaced by lodgepole and western hemlock more recently. In the Clackamas profile yellow pine superseded lodgepole for only a short period in the lower part of the profile, but remained significantly abundant during the rest of postglacial time. A sharp increase of lodgepole, to 49 per cent in the upper levels, suggests either fire or perhaps the influence of volcanic activity although no volcanic glass stratum was noted in the sedimentary column. Probably a yellow pine climax would have developed in much of the region within range of pollen dispersal to the sediments of this study in the Oregon Cascades except for the deposition of pumice at least twice during postglacial time.

WHITE PINE AND WHITEBARK PINE

The proportions of the pollen of white or whitebark pine are not known. The elevation of some of the bogs suggests that perhaps whitebark pine, an alpine species, is preponderantly represented. Next to lodgepole pine, these two pines are the most abundantly represented in some of the profiles. In Munson Valley, on the upper south slope of Crater Lake mountain, it supersedes lodgepole half-way up the profile and remains predominant to the present. This suggests modification of the soil and possibly an increase in moisture in later postglacial time. In the other profiles resting upon deep pumice white and whitebark pines are appreciably recorded, but remain more or less static for the time represented. Pollen of these species may have drifted down from higher life zones. In the Bend profile the recorded volcanic activity had little influence in changing the proportions of these species within range of pollen dispersal, as it

did in those of lodgepole and yellow pine. In the Rogue River and Prospect areas the trend of these two species has been generally constant, with a few sharp fluctuations that cannot be attributed to any definite cause. Their general greater abundance in the upper parts of the profiles may denote the influence of more moisture and lower temperatures in more recent time. The same trends for white and whitebark pine are recorded in the sedimentary columns from Clear and Clackamas Lakes. At no time were they predominant, and the minor fluctuations seem to be uninterpretable.

FIR

The total pollen proportions of the several species of fir indicate that as a group they have been next in importance to the pines on the crest and east slope of the Cascade Range. Represented in this group may be lowland white fir, white fir, alpine fir, noble fir, silver fir, red fir, and Shasta fir. The highest proportions of fir are recorded in the Rogue River profile and include chiefly white fir and alpine fir. The maximum attained is 38 per cent, the highest of any profile (fig. 103). This reflects the absence of the unfavorable influence of a deep pumice mantle, which has probably been responsible for its low porportions in profiles within the main pumice area. In other profiles the strongest representation of fir occurs in the Clear Lake, Clackamas Lake, Prospect, and Willamette Pass profiles which are located outside of the areas of maximum pumice deposition. In most profiles the general increase of fir in the upper strata suggests modification of the edaphic conditions as well as increased moisture and coolness since the middle postglacial warm, dry stage.

DOUGLAS FIR

Douglas fir has never been as abundant on the east slope of the Oregon Cascades as in the Willamette Valley and Puget Sound region. The drier climate and the edaphic conditions have probably been unfavorable for this species. It could not successfully compete with lodgepole in the pumice areas, nor with yellow pine beyond the pumice areas because of the dry climate. Its maximum proportions are attained in upper levels of the Rogue River and Prospect profiles which lies southwest of the principal pumice-covered areas. This suggests modification of the pumice-derived edaphic conditions and cooling of the climate and increased moisture in more recent time. In the other profiles it is recorded as having remained more or less constant with uninterpretable minor fluctuations.

WESTERN HEMLOCK

Hemlock has not flourished on the east slope of the Cascade Range in Oregon because of the dry climate and unfavorable pumice-derived edaphic conditions. It is recorded to an appreciable extent in only three profiles which are located beyond the pumice-covered areas and in moister situations. In the Clear and Clackamas Lake and Willamette Pass profiles hemlock reveals a general expanding trend from the bottom to the top, probably in response to amelioration of the environment since deglaciation, as well as a result of increased moisture and coolness in late postglacial time. Its highest proportion of 37 per cent is recorded in the Clackamas Lake profile at 0.7 meter (fig. 105).

MOUNTAIN HEMLOCK

The trend of mountain hemlock is similar to that of western hemlock, fir, and Douglas fir, in that it is more consistent and abundant in the upper levels of most profiles. This tends to support the evidence offered by the other species that there was an increase in moisture and coolness in later postglacial time which has persisted to the present. Its highest proportion of 12 per cent is attained in the Munson Valley profile which also is located in the Hudsonian life zone where mountain hemlock is an abundant species.

WESTERN LARCH

The present range of this species is less than halfway south in the Oregon Cascades, and it apparently has not extended any farther during postglacial time. Its most important record is that in the Bend profile where proportions as high as 33 per cent in the lower levels suggest severe and recurring fires just prior to the beginning of sedimentation (fig. 102).

A few other coniferous species that thrive in local abundance within the area represented by the sedimentary columns are western red cedar, incense cedar, Jeffrey pine, and Alaska cedar. A few pollen grains of the cedars were noted, but the pollen of these species is not durable and does not preserve well in organic sediments. Fresh pollen of Jeffrey pine has not been observed, but it is probable that some of the pollen listed as that of sugar or yellow pine includes some of the former. Its trends would be similar to those of yellow pine as its ecological requirements are similar.

Grasses, Composites, and Chenopods, the presence of whose pollen in quantity would suggest dryness and warmness, are poorly represented in all profiles. At no point or for no period of time are these groups sufficiently represented to suggest climatic changes.

Broadleaf arboreal species represented, but apparently with no systematic trends, are maple, alder, willow, and birch. The general increase in pollen of marsh and bog plants upward in the profiles indicates the usual course of hydrarch succession, and the development of the mature bog stage.

POSTGLACIAL VEGETATION OF THE NORTHERN GREAT BASIN

South central Oregon and the contiguous area of northern California are only sparsely forested. The forests are confined largely to the higher ridges and mountains where the precipitation is heavier. On lower slopes and poorer soil juniper scrub forest is predominant, while in the basins herbaceous vegetation is prevalent. All of the sedimentary beds in the northern Great Basin except one are situated in non-forested areas and at some distance from forests. Nevertheless, forest trees are predominantly represented in the pollen profiles up to the present, indicating that forests are over-represented in the sedimentary columns even though they be located in treeless zones.

Klamath Marsh rests directly upon Mount Mazama pumice, while Warner Lake and Chewaucan marshes contain an interbedded stratum of pumice from the same source. The four sections from Lower Klamath Lake do not seem to show any definite relationships to the eruption of Mount Mazama. Although Lower Klamath Lake was ponded in a Pleistocene basin, as were the other lakes of the northern Great Basin, no definite layer of volcanic glass was noted in the sedimentary columns. Possibly the main sheet of pumice underlies the sections examined. Volcanic glass, scattered throughout the sections, was probably carried by inflowing streams that head in the pumice areas in the Cascade Range to the northwest. Some of this glass may be from pre-Mount Mazama volcanic activity.

The warm, dry period between 8,000 and 4,000 years ago is well depicted in the pollen profiles from all sedimentary columns. Further evidence of drought is revealed by the occurrence of artifacts on the fossil lake bed of Lower Klamath Lake, exposed by wind and fire after drainage in 1917. This denotes the desiccation of Lower Klamath Lake along with the other Great Basin Lakes during the warm, dry stage (Allison, 1945; Antevs, 1940, 1945).

LODGEPOLE PINE

In only a single profile, one from Lower Klamath Lake, does the trend of lodgepole pine simulate those in the glaciated region (fig. 106). This is logical because none of the sediments are located within the glaciated region, and lodgepole was unable to benefit from the unstable physiographic conditions afforded by the retreating ice. There is no reason to assume that adjacent areas were not forested during the glacial period. In fact, the evidence suggests that forests were more abundant and widespread in the northern Great Basin during early postglacial time than at present. This indicates more favorable conditions, perhaps greater precipitation. In two of the sections from Lower Klamath Lake lodgepole remains

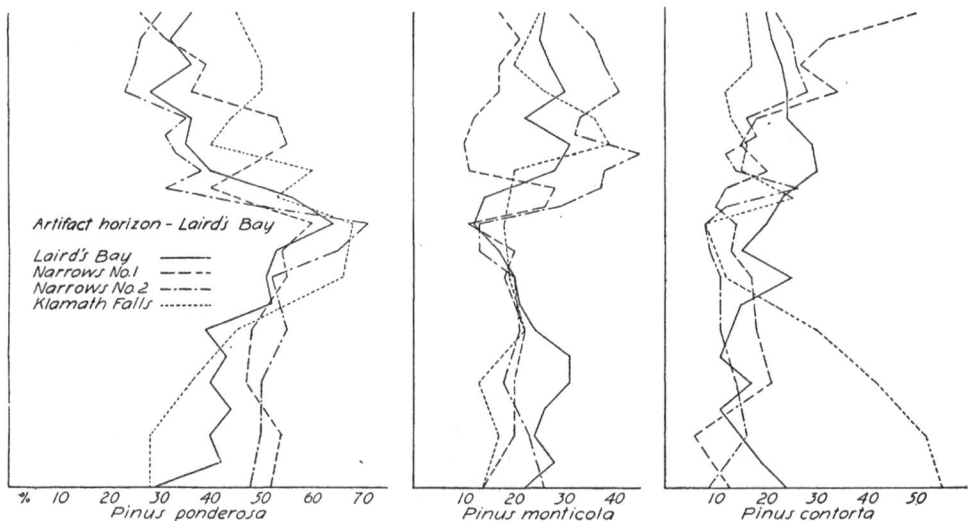

FIG. 106. Pollen profiles of western white, western yellow, and lodgepole pine in four sedimentary columns from the Klamath Lake region of northern California and south central Oregon. The influence of the warm, dry middle Postglacial is reflected by a decline of lodgepole pine in all columns. The artifact horizon in the Laird's Bay section occurs during the recorded dry interval.

more or less constant from bottom to top, while in the fourth it increases in the upper levels. The pollen profiles from Klamath Marsh reveal lodgepole in the minority in the lower and middle levels, and then it increases sharply in the upper levels to attain its maximum of 69 per cent at the top (fig. 107). The influence of the pumice mantle from Mount Mazama is not shown in the lower levels although the sediments rest upon pumice and lie in an area where the pumice is ten feet thick. Its increase to predominance in the upper levels suggests fire which favored an expansion of lodgepole. An area of lodgepole within the yellow pine forest immediately west and windward to the marsh explains its high proportions in the upper levels. The pollen profile of lodgepole pine from Chewaucan Marsh reveals a somewhat similar trend, with its maximum near the top (fig.

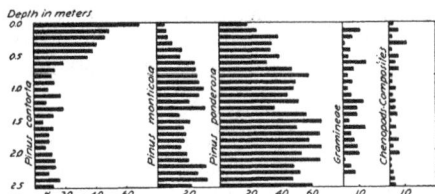

FIG. 108. Pollen diagram of sedimentary column from Chewaucan Marsh in south central Oregon. The warm, dry maximum is indicated by the high proportions of grass, Chenopods, and Composites.

108). Although a stratum of Mount Mazama pumice occurs in the Chewaucan section, it apparently was not thick enough in adjacent areas to influence forest succession as it did nearer Crater Lake. The influx of lodgepole near the top is probably due to its local expansion resulting from fire. A lodgepole pine stand covering about 4 townships 12 miles to the west of Chewaucan Marsh is probably the source of pollen. In the Warner Lake profile lodgepole is not abundant and attains its maximum of 25 per cent in the lowest level (fig. 109). Its lowest representation is evident during the warm, dry stage indicated above the Mount Mazama pumice. Apparently conditions in south central Oregon were not so favorable for lodgepole as in the glaciated region during early postglacial time, or in the central Oregon Cascades after the eruption of Mount Mazama.

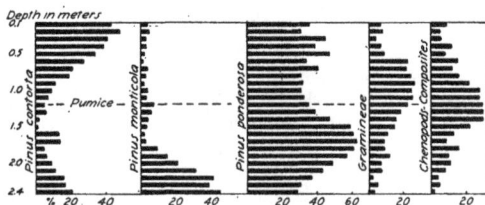

FIG. 107. Pollen diagram of sedimentary column on Mount Mazama pumice from Klamath Marsh. Postglacial warming and drying were well under way when the bottom sediments were deposited.

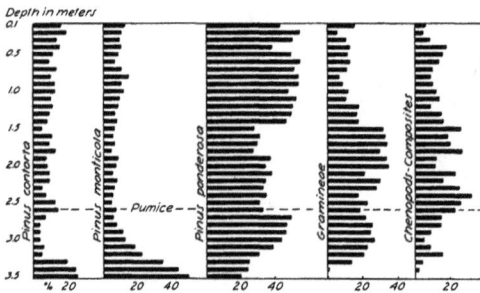

FIG. 109. Pollen diagram of sedimentary column from Warner Basin in south central Oregon. The high proportions of grass, Chenopods, and Composites strongly depict the warm, dry interval.

WESTERN WHITE PINE

The pollen listed as that of western white pine may include some of whitebark pine, but as both respond somewhat similarly to environmental changes their consideration collectively does not distort the picture. In all of the Lower Klamath Lake profiles white pine generally declines upward from the bottom to the middle of the profile, coincident with progressive warming and drying (fig. 106). Above the horizons deposited during the dry interval it slightly increases, reflecting the cooler and wetter climate of the past 4,000 years. In Klamath Marsh white pine maintains rather constant proportions to the upper few levels where it declines. The decrease in recent time is unexplainable unless fire destroyed upper slope types which were replaced with lodgepole in the mountains to the west. In Warner Lake and Chewaucan marshes white pine is most abundantly recorded in the lower levels and then decreases to a degree which is generally maintained throughout the balance of the profiles. As both of these sedimentary beds probably date back to early Postglacial, the trend of white pine reflects the cooler and moister conditions of that time, followed by warming and drying as postglacial time progressed.

WESTERN YELLOW PINE

In the Lower Klamath Lake sections an increase in yellow pine upward in all profiles to a maximum in the middle third followed by a decrease to or near the surface manifests its response to the warm, dry interval (fig. 106). In the Laird's Bay profile its maximum occurs just below the artifact horizon. An unconformity of unknown duration in this profile omits the record of adjacent vegetative succession during the maximum of the dry period. In the Klamath Marsh column yellow pine had attained a proportion of 60 per cent at the time of the earliest sediments (fig. 107). As these sediments lie upon Mount Mazama pumice, this indicates that desiccation had already progressed to a significant degree by the time

of the eruption. A maximum of 65 per cent several levels above the bottom and continued high proportions denote a protracted dry period, and then its slight decline for several levels in favor of grasses, Chenopods, and Composites, reflects increased warming and drying to a point that was unfavorable for yellow pine. After its decline to its minimum in response to the culmination of the drouth, it slightly increased, and then was somewhat displaced by an influx of lodgepole, probably of local occurrence immediately west and windward from the site of the sediments. In the Warner Lake and Chewaucan Marsh sections yellow pine slowly increases upward from the bottom, attaining its maximum and beginning to decline before the deposition of the pumice from the eruption of Mount Mazama. After the eruption continued warming and drying to a maximum caused it to be partially replaced by grasses, Chenopods, and Composites, and then after the culmination of the dry stage, it again expanded during the last 4,000 years. In the region of Chewaucan Marsh it was slightly displaced by a local expansion of lodgepole in a burned area several miles to the west.

GRASSES

The pollen record of grass in the northern Great Basin sections is extremely significant and strongly supports the occurrence of the warm, dry stage. In the Lower Klamath Lake profiles grass is sparsely represented, perhaps because of the unconformity in the sections caused by cessation of sedimentation when the lake dried up. In the Klamath Marsh column grass is appreciably represented and attains its maximum several levels above the bottom, or perhaps a few thousand years after the eruption of Mount Mazama (fig. 107). It remains abundant throughout the middle of the profiles, and then slightly declines toward the top, presumably in response to moister conditions in the past few thousand years. In the Warner Basin and Chewaucan Marsh sediments grass is recorded in low proportions near the bottom, but rapidly increases as postglacial warming and drying progressed. It attains its highest proportions after the eruption of Mount Mazama, denoting the culmination of the drouth sometime after this event. Its diminution in the upper third of both profiles expresses its response to a cooler and wetter climate in the past 4,000 years (figs. 108, 109).

CHENOPODS AND COMPOSITES

The pollen profiles of these two families, like those of grass, are extremely significant and strongly support the evidence for the warm, dry interval. In two of the Lower Klamath Lake profiles Chenopods are represented to a significant degree, but Composites are only sparsely and sporadically recorded. This may signify a gap in the column. In the Klamath

Marsh column Composites are better represented than Chenopods, but the persistence of yellow pine forests adjacent to the site of the sediments during all of post-Mount Mazama time has kept both of these groups to a minimum. In the Warner Basin and Chewaucan Marsh, however, they become very abundant during the middle third of the profiles above the Mount Mazama pumice stratum. Both groups show a decline in the upper third of the columns.

INTERPRETATION OF MISCELLANEOUS PROFILES

A few sedimentary columns have been obtained from bogs that lie outside of the areas represented by those groups already discussed. The indicated forest succession also seems to be somewhat different from that portrayed by the other groups, so it seems advisable to discuss briefly and interpret the vegetational succession as recorded in the profiles in relation to the present vegetation.

LAKE KACHESS

A sedimentary column has been obtained from a small bog located near the lower end of Lake Kachess in the central Cascade Range of Washington (fig. 110). The profile is only 4 meters thick and underlain with glacial drift, so it probably represents most or all of postglacial time. The volcanic ash stratum occurs at 3.2 meters, which is about the same relative stratigraphic level as in many other Washington profiles. The bog lies at an elevation of about 2,200 feet in the Canadian life zone. The upper limits of the timbered Arid Transition area lie a few miles to the east, and the Humid Transition area is several miles to the west. The windward-located Humid Transition is more strongly reflected than the Arid Transition area in the pollen profiles, although its nearest border is farther away. The adjacent forests are typical of the Canadian life zone, and the most abundant species are western hemlock, western white pine, Douglas fir, and lowland white, noble, and silver fir. Lodgepole pine is abundant on burned areas on the east and south slopes.

The recorded forest succession has been somewhat similar to that of the Puget Sound region, but with a few significant variations. Lodgepole was the pre-

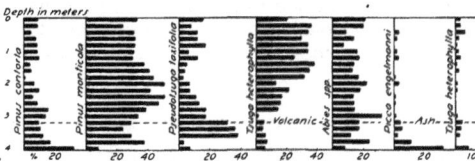

FIG. 110. Pollen diagram of the sedimentary column at Lake Kachess, in the Washington Cascades. The cooler and moister climate since the recorded volcanic activity is manifested by the rise of western hemlock.

dominant postglacial invader, and was rapidly replaced by Douglas fir and white pine, the former attaining its maximum below the volcanic ash stratum and then abruptly declining. White pine slightly diminished and then expanded to reach its peak of 51 per cent well above the ash layer, and then declined to the present. Western hemlock declined slightly up through the ash level, then abruptly increased to supersede white pine by a small margin near the top, and remained generally predominant to the present. The highest proportion attained by the balsam firs is 32 per cent just above the ash layer. The recorded trends of these species and groups suggest the influence of the xeric period, fire, and normal forest succession. The replacement of lodgepole by Douglas fir was normal forest succession which developed as the rigorous conditions left in the wake of the glacier were ameliorated. The concurrent expansion of white pine was also due to similar causes, while its slight decrease below the ash level may have resulted from the drying climate. Its rapid increase to its maximum may have been due to periodic light fire and increase in moisture. The rapid expansion of hemlock above the ash stratum indicates its climax development at the expense of white pine which is less tolerant. Since its pre-volcanic activity predominance, Douglas fir played a minor role, suggesting that the greater amount of precipitation and cooler climate than in the Puget Sound region permitted hemlock to expand, while occasional light fires were more favorable for white pine supremacy.

FARGER LAKE

Farger Lake is a large drained area near Crawford, Washington, about 10 miles north of Vancouver, Washington (map 1). The physiographic setting suggests that the original lake was ponded by fill deposited by glacial backwater from the Columbia River during the same time that the Willamette Valley was inundated toward the close of the last stage of the Wisconsin glacier. The site of the sediments lies south of the Puget Sound glacial boundaries and well below the limits of mountain glaciation to the east. The indicated postglacial forest succession in adjacent areas is strikingly different from that of either the Puget Sound region to the north or of the Willamette Valley to the south. Although Farger Lake is located in the Douglas fir forests, this species has not been predominant since the initial period of lodgepole pine invasion, except in more recent time, when it attained 39 per cent, its maximum (fig. 111). It is not recorded at all in the lower three levels, fluctuates between 0 and 7 per cent up to 1.2 meters, and then increases to its maximum at 0.6 meter. It then declines and is superseded by hemlock in the uppermost level. The most unusual trend is that recorded by Engelmann spruce. It shows 6 per cent at the bottom and

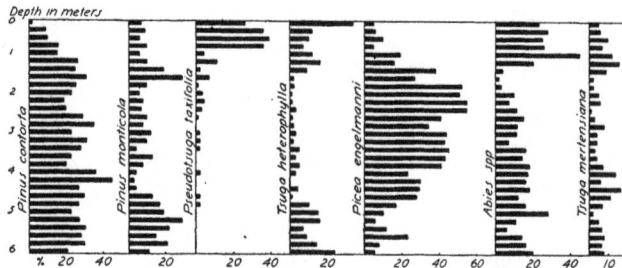

Fig. 111. Pollen diagram of the sedimentary columns at Farger Lake in south-western Washington. Although its age is uncertain, it is probably concurrent with those of the Willamette Valley. The high proportions of Engelmann spruce, which are uninterpretable, are the most significant feature. The increase in western hemlock near the top may be an expression of increased moisture in recent time.

then irregularly expands to 55 per cent at 2.4 meters. It maintains high proportions for four levels and then declines to the top. In no other sedimentary columns is Englemann spruce recorded to such high proportions, and its trend seems to be unexplainable either in terms of normal forest succession or climatic fluctuations. Its pollen may have been carried by wind and water from higher elevations in the Cascade Range to the east. Western hemlock, which in the Puget Sound region superseded Douglas fir in more recent time, is recorded to 24 per cent at the bottom, which is greater than usual, then generally diminishes to only 1 per cent at 4.6 meters, and maintains negligible proportions upward to 1.6 meters. From this horizon it increases to 34 per cent at the top where it is the most abundant species. The trends of fir are very similar to those of hemlock, both of which may reflect the influence of the warm, dry middle Postglacial. Lodgepole pine is predominant in the lower levels, is superseded by spruce, but remains more or less constant for a time and then declines in the upper levels to only 2 per cent at the top. It is possible that the sedimentary column does not represent all of postglacial time, but even so, this does not account for the unorthodox trends of the species concerned during the time represented.

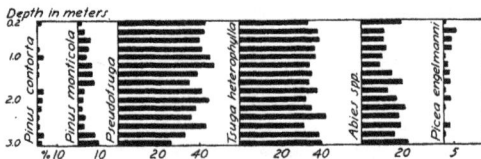

Fig. 112. Pollen diagram of the sedimentary column at Silver Lake in the Puget Trough of southwestern Washington. The age is unknown, but the column probably does not represent more than 4,000 years. The cool, moist climate is manifested by the high proportions of Douglas fir and western hemlock throughout.

SILVER LAKE

Silver Lake is a shallow lake located about 40 miles north of Farger Lake, in the Puget Lowland beyond the southern limits of glaciation (map 1). The lake is of unknown origin and age, but the pollen profiles suggest that it does not represent all of postglacial time. Volcanic glass in the lowest levels probably came from Mount St. Helens to the east, and may be of the same origin as that in Willamette Valley sedimentary columns to the south. In a three-meter

Fig. 113. Pollen diagram of one of the Cayuse Meadows sedimentary columns near Mount Adams, Washington. The influence of the Humid Transition life zone lying at lower elevations to the west and windward to the site of the sediments is revealed by high proportions of western hemlock throughout. The volcanic activity, probably local, is reflected by decline of hemlock and temporary rise of pine.

profile Douglas fir and western hemlock are coabundant throughout, with the former slightly predominant (fig. 112). Fir is next most abundant, while pine played a minor role throughout the time represented. The indicated forest succession is similar to that shown above the ash layer in most of the Puget Sound region profiles, and probably resulted from the same conditions.

CAYUSE MEADOWS

The sediments at this site have apparently accumulated in a depression caused by landslides in a glaciated valley. Cayuse Meadows is located about sixteen miles southwest of Mount Adams in the Cascade Range at an elevation of about 3,700 feet, in the

Canadian life zone (map 1) (Hansen, 1942e). The recorded forests for the time represented reflect the influence of the Humid Transition area at lower elevations to the west, and the timbered Arid Transition area at lower elevations to the east. Western hemlock is predominant throughout except at one level where volcanic ash occurs (fig. 113). The influence of the ash deposition is denoted by an influx of both yellow and lodgepole pine at the same level. It seems unlikely that the thin layer of ash in the vicinity was sufficient to change the composition of the forest. It may have caused a change in the preservative qualities of the sediments or a shift in the wind during the period of anthesis accompanying the volcanic activity. Since the wind normally blows from the west, the Humid Transition forests are perhaps over-represented. The source of the volcanic ash is unknown, but its relative stratigraphic position is similar to that in the other Washington profiles. One can hardly interpret the recorded forest succession in this sedimentary column upon the same chronological basis as those in the Puget Sound region or others in the Cascade Range.

ANTHONY LAKES

Two sedimentary columns from one of the Anthony Lakes in the Blue Mountains of northeastern Oregon, at an elevation of 7,000 feet, record the subalpine forest succession for an unknown period (Hansen, 1943a). The profiles, however, probably represent the time since the retreat of Pleistocene mountain glaciers. The chief competition has been between lodgepole and whitebark pine, with fire as probably the controlling factor. Whitebark pine was predominant in the lower half of each profile, but gives way to lodgepole pine supremacy in the upper levels (fig. 114). The influence of the Canadian life zone at lower elevations is shown by the abundance of lodgepole pine pollen where this species has persisted in great stands owing to fire. The yellow pine forests of the timbered Arid Transition area below the Canadian zone are represented in low proportions,

but an expansion about half-way up in the profiles may record the influence of the xeric period which may have caused an upward movement of the life zones. The location of the sediments at a subalpine elevation is reflected by low but consistent representation of alpine fir. A period of severe and recurring fire is denoted by an influx of western larch at a level about one third the way up the profiles. The most significant feature of these profiles is the further evidence of the existence of the postglacial xeric period as depicted by the expansion of yellow pine.

SUMMARY

The postglacial vegetational succession in the Pacific Northwest has been both regional and consistent in its response to environmental change. In each phytogeographic province the several profiles denote similar trends of plant succession, while over the entire region the climatic trends as interpreted from the pollen profiles are much the same although different species in each province are represented. Although the forest succession has been largely a response to the major, long-range climatic cycles, other factors have played a more important role in controlling plant succession in some areas.

In the Puget Lowland the pioneer, postglacial invader was lodgepole pine which took advantage of the rigorous conditions to colonize deglaciated terrain. It was rapidly replaced by Douglas fir as the environment was ameliorated. Western hemlock partially replaced Douglas fir, but its expansion to climax status was retarded by the warm, dry stage, and perhaps to some extent by fires. With the cooler and moister climate of late postglacial time, it expanded rapidly to become coabundant with Douglas fir, but fire continued to hinder its development to complete predominance.

In the coastal strip the normal plant succession which occurs with stabilization of the sand dune areas was interrupted from time to time by accelerated sand movement which permitted lodgepole pine periodically to regain predominance. In general, the succession has been from lodgepole pine to Sitka spruce and pine to spruce and hemlock, and in some cases to almost complete hemlock predominance. The presence of vast mature bogs in some areas in various stages of arboreal succession has tended to distort the picture of forest succession in adjacent areas.

In the Willamette Valley the pioneer lodgepole pine forests were rapidly replaced by Douglas fir which in turn was partially replaced by Oregon white oak during the warm, dry interval. As the climate became cooler and moister in more recent time, Douglas fir resumed predominance although oak has persisted abundantly to the present.

Pollen profiles from eastern Washington excellently portray the warm, dry stage between 8,000 and 4,000

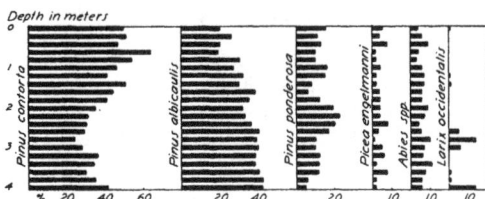

FIG. 114. Pollen diagram of one of the sedimentary columns at Anthony Lakes in the Blue Mountains of northeastern Oregon. The location of the bog in the Hudsonian life zone is indicated by the high proportions of whitebark pine. The repeated occurrence of fire is suggested by the abundance of lodgepole recorded. A single influx of larch indicates several fires occurring at short intervals.

years ago. The initial forests consisted largely of lodgepole, but with considerable western yellow pine. Yellow pine may have persisted south of the lodgepole pine belt near the ice front. As the climate warmed and the physiographic conditions left in the wake of the retreating ice became stabilized, yellow pine gained predominance, but continued warming and desiccation favored an influx of grasses, Chenopods, and Composites. With the advent of a moister and cooler climate yellow pine attained climax predominance which it has held to the present. The area held by grasses, Chenopods, and Composites during the dry interval has contracted considerably since the climatic maximum.

In northeastern Washington and northern Idaho the normal postglacial forest succession has been periodically interrupted by fires. The development of the forest climax consisting of western hemlock, western red cedar, and lowland white fir has been held in abeyance by these fires which have favored the persistence of subclimax species. The pioneer forests of lodgepole and western white pine have persisted in abundance during most of postglacial time. During certain periods severe fires occurring at frequent intervals favored an influx of western larch, probably much more extensively than the pollen proportions indicate. At other times stabilized conditions and the absence of fire permitted expansion of the climax dominants, but not to a predominant degree. In general, after the initial lodgepole invasion, western white pine has been the most abundant tree in this region during postglacial time. Its major expansion, favored by fire, was held in check until after the warm, dry stage. Appreciable proportions of grass in the lower horizons of some columns lend further support for the occurrence of the dry period more strongly pronounced farther south.

In the Oregon Cascades the forest succession has been influenced in some areas by the pumice mantle from Mount Mazama. Bogs resting upon this pumice reveal that lodgepole pine has been predominant to the present. In local areas other species have persisted, including yellow pine, western white pine, whitebark pine, Douglas fir, western hemlock, mountain hemlock, and Engelmann spruce.

North of the main pumice mantle the pioneer, arboreal invader was lodgepole pine which was partially replaced by western yellow pine as the climate became warmer and drier. Since the climatic maximum western hemlock, mountain hemlock, and western white pine have expanded, while yellow pine shows a slight decline. The persistence of lodgepole in appreciable proportions may probably be ascribed to occasional fires and perhaps local volcanic activity.

Sedimentary columns southwest of Crater Lake reveal that yellow pine has been the most abundant arboreal species since the eruption of Mount Mazama.

In this area the pumice is confined to the valley floors due to the pumice flows and the absence of air-borne pumice. The importance of the pumice for the abundance of lodgepole pine north and east of Crater Lake is shown by its low proportions in this area which has much the same climate.

Two sedimentary columns from the northern Great Basin of south central Oregon reveal that forests of white pine were predominant in the adjacent mountains during the Pleistocene, and persisted into the early Postglacial. The rapid expansion of yellow pine, however, reflects the warming and drying as postglacial time progressed. A marked influx of grasses, Chenopods, and Composites reflects the drouth of the middle Postglacial, and is followed by a slight resurgence of white pine in response to cooler and wetter climate of the past 4,000 years. In profiles from Lower Klamath Lake the warm, dry stage is reflected by an expansion of western yellow pine to predominance during or before paleic Indian occupied the lake bed. The absence of grasses, Chenopods, and Composites in appreciable proportions may be due to unconformities resulting from cessation of sedimentation during the drouth. In a section from Upper Klamath Marsh the climatic maximum is depicted by a limited influx of grasses, Chenopods, and Composites. As the section rests upon Mount Mazama pumice, only post-Mount Mazama plant succession is recorded.

In conclusion, it can be said that the pollen profiles of the Pacific Northwest reveal postglacial vegetational succession that is remarkably consistent in its implications with respect to climatic cycles. This regional climatic trend is manifested by the succession of many species of diverse ecological requirements, existing in several phytogeographic and climatic provinces.

POSTGLACIAL CLIMATE AND CHRONOLOGY

THE PROBLEM OF CORRELATION

In an earlier chapter an estimate was made of the length of postglacial time and the age of the sedimentary columns of this study. As was shown there, the thickness of the organic sediments is an inadequate criterion upon which to base other than relative chronological estimates. The range of thickness of the columns overlying the volcanic ash layer even within the same phytogeographic province demonstrates the great amount of variation in the depositional rate even under similar environmental conditions. The thickness of the columns in relation to their geographic position and to the probable relative time of deglaciation of their sites does not reveal any increase in the amount of sedimentation southward or even beyond the glacial boundaries. In fact the two thickest organic profiles in the glaciated area of the Puget Sound lowland are almost 200 miles apart; one near the southern glacial terminus and the other well within the glaciated region. The occurrence of a single volcanic ash stratum, apparently representing a single episode of volcanic activity in the Washington sedimentary columns provides a significant basis for chronological correlation. In Oregon profiles the presence of one or more pumice strata which can be differentiated as to source serves as a means of dividing postglacial time into two or more units upon at least a relative basis. These time markers correlate with the pollen-analytical chronology as well as with the data derived from non-botanical sources, including archaeology, glaciology, geology, climatology, human history, and others, providing a general scale for postglacial time in the Pacific Northwest. The linking of human prehistory with the pollen-analytical sequence in Europe has made considerable progress. · In North America the correlation of the evidence of early man with the pollen records has had hardly more than a beginning, but the results, though meager, offer some encouragement. Unfortunately, the oldest records of man are found in the more arid regions of the continent where hydrarch succession and the development of sediments holding a pollen record have been scarce.

THE BLYTT–SERNANDER SCHEME

A well-known and widely accepted climatic sequence for the Postglacial of Scandinavia is the Blytt-Sernander scheme. This scheme was formulated by Blytt (1881) upon the basis of typological succession in peat bogs and upon peat stratigraphy in Norway. Sernander (1894, 1908, 1910) simplified Blytt's original scheme, and since then it has been known as the Blytt-Sernander scheme. It has been applied to other parts of northern Europe, and has been correlated with absolute chronology based upon varved clay studies. Beginning with the initial retreat of the last ice sheet in Sweden, the first stage is a cold Arctic, followed by a cold to cool sub-Arctic. These stages comprise most of the younger late-glacial. The third stage is the boreal which was cool to warm and dry, succeeded by the warm and moist Atlantic stage during which the temperature maximum, was attained. The fifth stage, the sub-boreal was warm and dry and is known as the xerothermic period (Sears, 1942). The final stage, the sub-Atlantic marks a return to cooler and moister conditions, considered by some as climatic deterioration. The last four stages, comprising practically all of the Postglacial, represent about 9,000 years (table 8).

Correlated with the Blytt-Sernander sequence are the several recent geologic stages of the Baltic Basin (Osborn, 1922; Granlund, 1936; Movius, 1942). A freshwater lake occupied the basin during the Arctic and about half of the sub-Arctic. This was replaced by the cold Yoldia brackish-water sea for about 600 years during the latter half of the sub-Arctic. The freshwater Ancylus Lake existed during the balance of the sub-Arctic and all of the boreal stage. The warm Litorina Sea followed and prevailed during the warm, moist Atlantic stage. The sub-boreal and sub-Atlantic stages were represented by the Limnaea or Baltic Sea, which has occupied the basin to the present (table 8). ·

THE DE GEER–LIDÉN CHRONOLOGY

The application of De Geer's varved clay chronology to the Blytt-Sernander climatic sequence has provided an approximate absolute time scale for postglacial time in Europe (De Geer, 1910, 1940; Antevs, 1925a). According to this chronology, De Geer considered that the ice age in Sweden terminated with the bisection of the ice remnant west of Ragunda in northern Sweden. This bisection has been dated at 6839 B.C. by Lidén (1938), which makes the Scandanavian Postglacial about 8,800 years long. Applying this chronology to the Blytt-Sernander climatic sequence, the Arctic in Sweden began with the initial retreat of the last ice sheet in Sweden about 15,000 years ago and lasted until 11,000 years ago. The sub-Arctic, comprising the rest of the younger lateglacial, followed and continued until about 9,100 years ago. The boreal began near the close of the lateglacial and prevailed up to about 7,500 years ago. During the next period, the Atlantic, which is defined as having persisted until 4,500 years ago, the Pleistocene ice sheet entirely melted. The sub-boreal persisted until about 2,600 years ago, and the final period, the sub-Atlantic, continues to the present (table 8).

THE VON POST SCHEME

Von Post (1930), upon the basis of pollen analytical data from bogs of northern Europe, has stressed the most important change in climate by dividing the Postglacial into three stages. The first was one of increasing warmth, the second stage was one of maximum warmth, and the final stage was one of decreasing warmth. Von Post (1933) states that during and just after glacial retreat the temperature gradually rose and attained its maximum between 7,000 and 6,000 years ago. Pollen profiles show that during the · past 4,000 years there was a decline of forest trees requiring long and warm growing seasons, suggesting a definite decrease in temperature. Von Post's middle stage of maximum warmth approximates the later two-thirds of the Atlantic period. At the same time von Post (1933), in his table, includes the boreal, Atlantic, and sub-boreal stages of the Blytt-Sernander scheme in the "Age of Warmth." Von Post (1933) further states that the Blytt-Sernander scheme is fundamentally correct, but that research during the last decades has considerably modified our views and has also shown that there were more climatic stages

TABLE 8

	LATE-GLACIAL AND POSTGLACIAL CHRONOLOGY AND CLIMATE IN SWEDEN						LATE-GLACIAL AND POSTGLACIAL CHRONOLOGY, CLIMATE, AND VEGETATION IN EASTERN NORTH AMERICA		
De Geer Lidén	Baltic Basin	Blytt-Sernander	von Post	Granlund		Antevs	CLIMATE and VEGETATION Sears, Deevey	GLACIAL CHRONOLOGY Antevs	CLIMATIC AGES Antevs
Years									
1,000		Sub-Atlantic	Decreasing temperature	Cold, wet		LATE	Sub-Atlantic / Cooler-Moister / Oak-chestnut-spruce		Sub-Atlantic
2,000	POSTGLACIAL — LIMNAEA SEA	Cooler Moister					(Oak-beech)		
3,000		Sub-boreal		Warm summers (Trapa)		POSTGLACIAL	Sub-boreal warm-dry / Temperature maximum		
4,000		Warm-dry					Oak-hickory		
5,000		Atlantic	Maximum warmth	Postglacial age of warmth	Mild winters (Cladium)	MIDDLE	Atlantic / Warm-moist		Sub-boreal
6,000	LITORINA SEA	Warm					Oak-hemlock		
7,000		Moist					(Oak-beech)		
8,000	ANCYLUS LAKE	Boreal Cool to warm dry				EARLY	Boreal / Warmer than Pre-boreal but cool / Pine		
9,000	FINI-GLACIAL						Pre-boreal Cool-Moist Spruce-fir		
10,000	Yoldia Sea	Sub-Arctic Cool						Ice-border in region of Cochrane, Ont.	
11,000	BALTIC ICE LAKE						Hudsonian		Atlantic
12,000	GOTHIGLACIAL	Arctic Cold						Rapid retreat of ice	
13,000									
14,000							Pre-Hudsonian		
15,000								Ice-border at Mattawa, Ont.	
20,000									Boreal
25,000								(Vashon) Mankato Maximum / Cary Maximum	Pre-boreal
30,000									

of this kind than had previously seemed probable. Godwin (1940) interprets the pollen profiles of England and Wales upon the basis of the von Post sequence, while Fuller (1935) applies the same sequence to the pollen record of the Lake Michigan region.

Granlund (1932, 1936), upon the basis of peat stratigraphy, presents evidence for several abrupt changes from dry to wet conditions in Sweden, which he calls "recurrence surfaces." They are dated as follows:·recurrence surface I, A.D. 1200; II, A.D. 400; III, 600 B.C. (boundary horizon between the sub-boreal and sub-Atlantic); IV, 1200 B.C.; V, 2300 B.C.; VI, 2900 B.C.; and VII, about 3700 B.C. It does not seem probable that peat stratigraphy offers so reliable evidence for regional climatic fluctuation as do pollen profiles. It is to be noted that peat horizons indicating climatic changes are not consistently found in all bogs even within limited areas (Andersson, 1892).

ANTEVS' SCHEME

Antevs (1931, 1933) believes that long-distance correlations, especially trans-oceanic, can be made only upon the basis of great, slow changes in the summer temperature. Variations in precipitation, being zonal or regional, cannot be used with assurance or at all. Antevs uses temperature variations for the division of late-glacial and postglacial time. The postglacial age of distinctly higher temperature is designated as the middle Postglacial. This stage is dated in Sweden and Denmark from 6000 to 2000 (possibly 1000) B.C., and is used as a starting point in correlation and in application of the Swedish varve chronology to North America. Other starting points employed are the oscillations of the ice border at Cochrane, Ontario (Antevs, 1928), and the central Swedish moraines—the Salpausselkas of southern Finland, which may be of the same age, though, because of local conditions, the former represent a longer time (Antevs, 1931). The Salpausselkas were formed during the years from 1910 to 1250 before the zero year (Sauramo, 1929), which is 6840 B.C. (Lidén, 1938). This means they were formed from 10,690 to 10,030 years ago. The date of the Cochrane oscillations of approximately 11,500 to 9,500 years is a useful time marker.

EASTERN NORTH AMERICAN CLIMATE AND CHRONOLOGY

In late-glacial and postglacial pollen profiles of eastern North America, Sears (1941, 1942, 1942a) and Deevey (1943, 1944) see both the climatic and chronologic sequences of the Blytt-Sernander scheme. They have correlated their climatic stages with the equivalent ones in Sweden, and have tentatively applied the dates of the periods in Sweden to the stages in North America. Deglaciation of southernmost Sweden began some 15,000 years ago, and the Stockholm region was freed of ice 10,000 years ago. Sears'

and Deevey's direct time correlation thus implies that the pollen profiles of eastern North America (largely in Ohio and Connecticut) represent only about 10,000 years. Deevey's profiles occur on Tazewell drift (MacClintock and Apfel, 1944) and many of Sears' profiles lie on Cary drift. Such a correlation does not take into consideration that these areas became ice-free approximately 30,000 years ago. Both Sears and Deevey have suggested that the kettles in which the organic sediments have accumulated were occupied by dead ice for some time after the glaciers had disappeared. It seems hardly possible, however, that such ice, in relatively small and shallow basins, could have persisted for more than a few centuries. The existence of tundra conditions for some time after deglaciation has also been suggested, but the present occurrence of forests on and near glaciers in Alaska, and well above glacier termini on our higher western mountains tends to minimize this possibility. If tundra had persisted for a long time after deglaciation, a record of such vegetation should be present in the lower sediments deposited before the advance of the forests. Bog plants are largely boreal, and many of them thrive on the tundra and sub-tundra of North America today. There seems to be no reason why such plants could not have thrived on recently deglaciated terrain or even contributed to the accumulating sediments in ponds and lakes. Thus, if boreal forests did closely follow the retreating ice, and the sedimentary columns represent only 10,000 years and yet record boreal forests in their lowest levels, then such forests must have persisted for 20,000 years or more after deglaciation before they were replaced by pine. Sears' and Deevey's climatic and chronologic correlations are admittedly tentative. They are significant, however, because they constitute a beginning of inter-hemispheric correlations, which may be refuted or substantiated by further pollen analyses in other regions and by work in other fields, the results of which may furnish further evidence for climatic cycles.

Smith (1940) has attempted to correlate 148 eastern North American pollen profiles with the Blytt-Sernander sequence, and believes that all periods beginning with the pre-boreal are evident in most profiles. He suggests that the spruce maximum is to be correlated with the oscillations of the last Wisconsin ice sheet at Cochrane, Ontario, between 9,500 and 11,500 years ago (Antevs, 1939).

Antevs [5] would extend these American vegetational and climatic sequences over a much longer period of time than do Sears and Deevey. He believes that as the last deglaciation in eastern North America and in Europe probably required more than 30,000 years and was accomplished only some 5,000 years ago, the duration of the combined late-glacial and postglacial

[5] Personal communication.

ages ranges from almost 40,000 at the peripheries of the glaciated areas to 5,000 years at their centers. Therefore, when the climates in many glaciated regions show an essentially similar evolution since the release from the last ice sheets, they have obviously passed through the stages in different lengths of time. As a consequence, direct time correlations of the late-glacial and postglacial climatic periods can be made only when the regions became ice-free at about the same time. In other cases, correlations of at least the early comparable periods have to be made on a sliding time scale. For instance, since northern Vermont became ice-free about 5,000 years later than did Connecticut, the pre-boreal age there began 5,000 years later; and it also ended later, perhaps 3,000 to 4,000 years later, than in Connecticut. When the ice border had withdrawn to the region south of James Bay, the time lag may have been so reduced that the sub-boreal began at about the same time in northern Vermont as in Connecticut.

Since tundra or forest vegetation followed the retreating ice borders, and since the peat and pollen profiles in Ohio and Connecticut occur in regions that became exposed to vegetation about 30,000 years ago, they may represent this length of time rather than 10,000 years. Thus, the forest sequence and the climatic history worked out by Sears and Deevey should perhaps be distributed over the past 30,000 years (table 8). As a first step in this dating, Antevs would directly correlate that postglacial age in North America during which the temperature maximum as its middle part, enjoyed higher temperature than now prevails. In this country Sears' and Deevey's sub-boreal was the warmest; and Antevs would date this period (together with preceding and/or following centuries that might also prove to have been warmer) at 6000–2000 B.C. According to this correlation and dating, the warm and dry American sub-boreal would be practically contemporaneous with the warm and moist Swedish Atlantic, although perhaps precipitation should not be considered in trans-oceanic correlations. Antevs has classified the warmer age as the middle Postglacial (Postpluvial) (table 8).

The American warm and moist Atlantic, characterized by forests of beech, oak, and hemlock, may have begun some 15,000 years ago, contemporaneous with a quickening of the ice retreat north of Mattawa, Ontario, and the inception of a period of warm summers according to the Spitaler data (table 8). The pre-boreal forests, represented by fir and spruce, would be assigned to the time immediately following deglaciation, or in Connecticut and central Ohio about 30,000 to 20,000 years ago. The secondary pre-boreal spruce maximum in some of the profiles could then be truly correlated with the Mankato-Valders Maximum.

Other pollen profiles from the Great Lakes region do not clearly show much correlation with the Blytt-Sernander scheme or the North American sequence as interpreted by Sears. At least the investigators of bogs in this region do not read any such meaning into them. In fact some of the more active workers have been loath to interpret any postglacial climatic changes upon the basis of the forest succession indicated in the pollen profiles (Wilson, 1938; Potzger and Friesner, 1939) except a general amelioration as the ice sheets waned. More recent work and interpretations by these workers have provided some basis for postulating late-glacial and postglacial climatic fluctuations. There is some evidence in pollen profiles from northern Wisconsin, Minnesota, and western Ontario for a single xerothermic period (Potzger, 1942; Potzger and Richards, 1942; Wilson and Webster, 1942). The absence of two dry periods to correlate with Sears' sequence is suggested by Wilson (1944) as due to the fact that bogs providing the data for the former sequence are located on Cary drift, while the more northern deposits rest on the younger Mankato-Valders drift. Wilson and Webster also see a sequence in these same profiles that they tentatively consider as supporting von Post's (1930) scheme. It is to be noted that the sub-boreal and the sub-Atlantic stages of eastern North America are climatically equivalent to the second and third European stages of von Post, but direct trans-Atlantic chronological correlations can hardly be presumed if North American profiles represent 25,000 years as compared with 9,000 to 10,000 years in Europe. Work by Wilson in south central Ohio, as yet unpublished, agrees with Sears' fivefold division of post-glacial climate (Wilson, 1944). Wilson further states, however, that in southern Ohio the pollen record began with the retreat of the Tazewell ice, probably about 40,000 years ago. Pollen analyses of bogs in and near the Driftless Area of southwestern Wisconsin reveal evidence for a climatic sequence similar to that of von Post (Hansen, 1937, 1939c). The middle and last stages may be chronologically equivalent to those of von Post, but the initial period of warming was probably several times longer than the first stage of the von Post scheme. Some two hundred peat profiles in the Great Lakes region and New England have been analyzed by various workers, but the discussion of detailed interpretive results here would require too much space. In general, many of these pollen profiles seem to indicate that there was at least one stage of postglacial warming, followed by cooler and moister conditions. Whether or not this warm stage was one of alternating dryness and humidity may be determined with more certainty as more peat columns are analyzed and forest succession better understood and defined.

PACIFIC NORTHWEST CHRONOLOGY

The sedimentary columns of the Pacific Northwest that rest upon glacial drift or its chronological equivalent are estimated to represent an average of 18,000 to 20,000 years. This period is arbitrarily called the Postglacial in interpreting the Pacific Northwest pollen profiles, regardless of glacial movements and time of final dissipation of the ice sheets in various parts of the continent. Upon the basis of the recorded forest succession this time is roughly divided into four periods. Period I is set as the time from the last glacial maximum to 15,000 years ago. Period II is dated from 15,000 to 8,000 years ago, and marks a time of increasing warmth and dryness. During this stage perhaps about 10,000 years ago, the temperature reached a level similar to that of today. Period III endured from 8,000 to 4,000 years ago and marks the stage of maximum warmth and dryness. This stage is contemporaneous with Antevs' warm, dry middle Postglacial. The final stage, Period IV, saw a return to a cooler and moister climate which has persisted in general to the present. The volcanic ash in the Washington sections is dated at about 6,000 years, or about the middle of the warm, dry period (table 9).

The eruption of Mount Mazama is dated at about 10,000 years ago, and any bogs resting upon its pumice cannot be older. Other sedimentary columns in the Oregon Cascades that rest upon glacial drift are estimated to represent about 15,000 years, there presumably having been a slight lag in Pleistocene mountain deglaciation. The profiles from Lower Klamath Lake may not be as old, and moreover, contain an unconformity of unknown magnitude due to cessation of sedimentation for a period of time during the warm, dry middle postglacial.

EARLY POSTGLACIAL TIME IN THE PACIFIC NORTHWEST

In the Pacific Northwest the postglacial time represented by the sedimentary columns resting upon glacial drift can be divided into two periods, those before and after the volcanic activity recorded by the volcanic ash stratum. The span of time represented by the sediments overlying the ash layer is the same for all profiles, excepting disparities due to possible unconformities. The sedimentary columns underlying the ash may vary somewhat in total age, but there are two definite starting points for chronological correlation, the top of the column (the present), and the ash stratum. The time lapse between deglaciation and initial pollen-bearing sedimentation probably varies somewhat for each profile. The close correlation of the pollen profiles in the lower levels, however, suggests that no great time differential is involved.

The absence of pollen of boreal species in significant proportions in the bottom layers suggests the absence of tundra conditions, such as may have existed upon deglaciation in eastern North America (Antevs, 1922, Hollick, 1931). Neither is there evidence for the existence of timberline arboreal species near the sites of the sediments during early postglacial time. If they did exist in any abundance, sedimentation did not begin until these species had migrated to higher elevations. The striking predominance of lodgepole pine during early postglacial time is evidence for a rigorous environment, particularly as to edaphic and physiographic conditions but not necessarily as to climate.

The absence of a boreal continental climate such as existed farther inland was undoubtedly due to the proximity of the Pacific Ocean which presumably moderated the climate then as it does today, notwithstanding a possible difference of a few degrees in absolute temperature. The early postglacial climatic conditions west of the Cascades were probably little different from those near the margins of the present-day Alaskan glaciers. There, forests exist close to the ice front (Cooper, 1942) and even upon dead masses of ice (Tarr, 1908; Washburn, 1935). With the exception of mountain hemlock these species are the same that dominate the forests of the Puget Lowland and coastal strip today. The presence of a nearby ice front evidently did not cause a boreal or even sub-boreal climate in the Puget Lowland during early postglacial time. Even today nonboreal forests occupy the ridges between valley glaciers on the higher western mountains, well above the termini of these glaciers. The climatic conditions now required to maintain these comparatively small valley glaciers emanating from small catch-basins certainly must be almost as severe as those necessary to develop and maintain a lobe of the continental ice sheet that extended hundreds of miles from its center of accumulation, and to maintain a climate permitting its persistence. That the climate was cooler and moister during the early Postglacial than at any time since is suggested by the highest proportions of western white pine in lower levels coincident with the maximum of lodgepole pine.

East of the Cascade Range the early postglacial climate was undoubtedly colder just as it is more nearly continental today, although no tundra species or boreal forests of present timberline species are recorded in the lower levels to any significant degree. It is possible, however, that pollen-bearing sediments were not deposited during early postglacial time, or if they were they may have been removed by shifting streams. Sedimentary sites lying beyond the boundaries of glaciation, however, could not have been occupied by dead ice, a reason advanced for the absence of early postglacial sedimentation in kettle ponds elsewhere. These extra-glacial sites likewise do not reveal in the lowest levels of their sediments any evidence of tundra flora or of timberline forests. The pollen profiles of sections located in the treeless areas of eastern Washington and Oregon indicate that forests were most widespread in early postglacial time, reflecting a cooler and wetter climate persisting from the Pleistocene.

TABLE 9

CLIMATE AND CHRONOLOGY IN THE PACIFIC NORTHWEST   LATE GLACIAL AND POSTGLACIAL

Years ago	CLIMATE OF THE GREAT BASIN ANTEVS	ASH and PUMICE	CLIMATE PACIFIC NORTHWEST	PUGET SOUND	EASTERN WASHINGTON	NORTHERN IDAHO
1000		•		HEMLOCK PREDOMINANCE AND MAXIMUM Douglas fir slight decline	YELLOW PINE PREDOMINANCE AND MAXIMUM Grass static	Slight rise of climax dominants, hemlock and fir Lodgepole slight rise
2000	COOLER MOISTER		PERIOD IV COOLER-MOISTER	White pine and lodgepole static	Chenopods-Composites static White pine slight rise	Larch influx
3000				Fir static	Lodgepole static Yellow pine rapid rise	WHITE PINE PREDOMINANCE AND MAXIMUM Larch influx
4000		Devil's Hill pumice		Hemlock rapid rise	Grass static	Lodgepole decline to minimum
5000	WARMER and DRYER THAN PRESENT	Willamette Valley pumice	PERIOD III MAXIMUM WARMTH AND DRYNESS	Douglas fir decline White pine slight rise Lodgepole pine static	Chenopods-Composites decline Lodgepole static	White pine expansion
6000		Washington volcanic ash			CHENOPOD-COMPOSITE MAXIMUM	
7000				Douglas fir slight decline White pine minimum	Grass rapid decline Chenopod-Composite rapid expansion	Lodgepole slight rise
8000				Lodgepole static	Yellow pine slow rise GRASS MAXIMUM	GRASS MAXIMUM
9000	INCREASING WARMTH AND DRYNESS			Hemlock slight expansion Fir static	Chenopod-Composite slight rise Yellow pine static	White pine static
10000		Mount Mazama pumice	PERIOD II INCREASING WARMTH AND DRYNESS	Douglas fir predominance	Grass rapid rise	
11000						
12000				DOUGLAS FIR PREDOMINANCE AND MAXIMUM Fir static	Yellow pine slow rise White pine static	Grass slight rise
13000	RISING TEMPERATURE			Hemlock expansion arrested Lodgepole rapid decline	Chenopods-Composites static Lodgepole continued decline	Lodgepole decline White pine slight rise
14000	DECREASING MOISTURE			DOUGLAS FIR RAPID RISE TO SUPERSEDE LODGEPOLE PINE	GRASS RAPID EXPANSION	LODGEPOLE PINE PREDOMINANCE AND MAXIMUM
15000			PERIOD I COOL MOIST	Hemlock slow rise Lodgepole rapid decline	Grass expansion	
16000				Douglas fir expansion	Chenopods-Composites static Yellow pine rise	
17000	SUBSIDING LAKES			White pine maximum LODGEPOLE PREDOMINANCE AND MAXIMUM	White pine maximum LODGEPOLE PREDOMINANCE AND MAXIMUM	
18-20000						

Side labels (Climate of the Great Basin Antevs column): POSTGLACIAL — LATEGLACIAL; LATE, MIDDLE, EARLY — YOUNGER, MIDDLE; POSTPLUVIAL (LATE, MIDDLE, EARLY); PROVO PLUVIAL MAXIMUM ABOUT 23,000 YEARS AGO

Along the Pacific Coast those profiles that possibly represent all of postglacial time reveal that the earliest forests consisted of species of the same identity and of about the same proportions as those of the climax forest of today. As most of this area lies well beyond the glacial boundaries and under the dominating climatic influence of the adjacent ocean, the former presence of forests of modern aspect eliminates any possible interpretation of a climate essentially different from that of the present. In those montane profiles that presumably record most of postglacial time the earliest forests were likewise little different in composition from those of today. There is no evidence that timberline species such as mountain hemlock, alpine fir, whitebark pine, alpine larch, and Alaska cedar existed in greater abundance at lower elevations at that time than now. In the Willamette Valley, where the profiles are probably approximately equivalent in time to those of the Puget Lowland in Washington, no timberline species are recorded, while

TABLE 9—(Continued)

VEGETATION IN THE PACIFIC NORTHWEST POST–MOUNT MAZAMA FOREST SUCCESSION

WILLAMETTE VALLEY	OREGON CASCADES	TUMALO LAKE NEAR BEND, OREGON	CHEWAUCAN MARSH WARNER BASIN	LOWER KLAMATH LAKE	ON THICK PUMICE	ON THIN PUMICE	Years ago
DOUGLAS FIR MAXIMUM	Hemlock slight rise	Lodgepole decline	Lodgepole increase in local areas	All species of pine generally static	Yellow pine static	Douglas fir and fir increase	
Hemlock static to slight rise	Yellow pine slight decline	Yellow pine White pine Whitebark pine static		Slight decline of white pine	Slight increase of Mt. Hemlock White pine Whitebark pine	Yellow pine static	1000
Fir slight rise	Lodgepole increase		Yellow pine increase	Yellow and lodgepole pine static			2000
Elimination of lodgepole and spruce	White pine Whitebark pine Douglas fir generally static	Lodgepole increase				Lodgepole, white, and whitebark pine static	
Oak decline		Yellow pine decline	Grass Chenopods-Composites decline	Rise of white pine to maximum	LODGEPOLE PREDOMINANT AND STATIC		3000
		Devil's Hill		Yellow pine decline		Yellow pine decline	
	YELLOW PINE MAXIMUM	pumice		YELLOW PINE	LODGEPOLE PREDOMINANT AND STATIC		4000
Pumice		Yellow pine expansion	Yellow pine increase	Artifact horizon		White and whitebark pine static	5000
OAK MAXIMUM	Lodgepole pine White pine Whitebark pine static		CHENOPOD-COMPOSITE MAXIMUM	MAXIMUM			
Lodgepole, fir, spruce continue to decline		White and whitebark pine static	Yellow pine continued decline	Yellow pine expansion	Other species generally static with minor fluctuations	YELLOW PINE MAXIMUM	6000
Hemlock decline	Douglas fir abundant	Lodgepole decline	GRASS MAXIMUM	White pine decline		Lodgepole static	7000
Oak rapid rise		Yellow pine increase LODGEPOLE MAXIMUM Yellow pine minimum	Grass Chenopods-Composites continued expansion	Lodgepole, white, whitebark pine irregular with non-correlative fluctuations	LODGEPOLE PREDOMINANT AND STATIC Other species static LODGEPOLE PREDOMINANCE	Lodgepole decline YELLOW PINE PREDOMINANCE	8000
	Lodgepole decline						9000
Douglas fir expansion retarded		Mount Mazama	Mount Mazama		Mount Mazama	Mount Mazama	10000
		pumice	pumice		pumice	pumice	
DOUGLAS FIR RAPID EXPANSION		YELLOW PINE MAXIMUM	Chenopods-Composites expansion				11000
		Yellow pine expansion	Grass rapid expansion				
Lodgepole, fir, spruce decline	Yellow pine expansion		Lodgepole and white pine decline	?			12000
Oak absent		Lodgepole; larch decline	Yellow pine expansion				13000
Hemlock static	LODGEPOLE PREDOMINANCE AND MAXIMUM	LODGEPOLE AND LARCH PREDOMINANCE	WHITE PINE PREDOMINANCE AND MAXIMUM				
DOUGLAS FIR SLOW RISE							14000
							15000
Lodgepole decline							
Douglas fir, hemlock scarce							16000
Spruce-fir maximum							
LODGEPOLE PREDOMINANCE AND MAXIMUM							17000
							18–20000

Canadian zone species have only a limited representation. White pine trends suggest a cooler initial stage than that which followed.

In general there were no extensive early postglacial forests in the Pacific Northwest that were the climatic equivalent of the late-glacial spruce-fir forests of eastern North America, or of the much later birch-pine forests of Europe possibly indicative of a somewhat similar climate. The pioneer forests of lodgepole and white pine, however, may have been chronologically equivalent to those of pine in eastern North America, during the boreal stage as redated by Antevs (table 8). With the exception of the coastal strip the early postglacial climate of the Pacific Northwest was probably cooler than it has been at any time since, while, of the entire area, the climate of the Columbia Basin was perhaps most nearly like the pre-boreal climate of eastern North America and Europe.

MIDDLE POSTGLACIAL TIME IN THE PACIFIC NORTHWEST

GENERAL CONSIDERATIONS

There is strong evidence in the pollen profiles for a warm, dry interval in the Pacific Northwest, that was concurrent with Antevs' middle postglacial stage of maximum warmth in the Great Basin. The stratigraphic position of the Washington volcanic ash in relation to the recorded warm stage, and the climatic sequence of the Great Basin and others parts of North America and of northern Europe, suggest that this period of maximum temperatures may well have occurred between 8,000 to 4,000 years ago (table 9).

In interpreting the replacement of the pioneer forests of lodgepole and western white pine by other species, as well as subsequent forest succession, the possibility of normal forest succession must be considered. Also alteration of the forest composition as a result of insect and fungus disease, wind, fire, landslides, snowslides, and vulcanism, although localized in area, may be reflected in the pollen profiles. Such fluctuations may simulate those due strictly to climatic changes, and these causal factors must also be considered as alternatives to the response to a change of climate or to normal plant succession due to moderation and stabilization of the environment. If the indicated trends are consistent and systematic over wide areas with evidence for contemporaneity, the error in interpreting such trends in terms of climate probably is largely eliminated, inasmuch as, of these several factors, climate is the most nearly regional. This is especially true in the Pacific Northwest where the climatic trends in the past are indicated even in areas which today have different local climates and support different species of varying ecological requirements. Despite these differences a measure of unity in the trends is common to all. Only major regional changes of climate afford an adequate explanation of this fact.

COASTAL STRIP

On the coastal strip adjacent to the Pacific Ocean there is little indication of a climate drier and warmer than the present at any time during the Postglacial. The marine influence has moderated the climate and the available moisture has probably never been a limiting factor. The principal cause of change in forest composition has been the periodic shifting of sand, disrupting succession toward the climax vegetation or destroying the existent climax forests. Succession on mature bogs is also reflected in the pollen profiles. An influx of spruce recorded near the middle of each of nine profiles may reflect the influence of increased moderate sand movement for a brief time (fig. 95). This in turn may have been contemporaneous with the drier trend recorded in peat profiles located farther inland. The limited probable

age of most of the coast sedimentary columns, however, and the high position of the spruce expansion recorded therein suggest that it may have occurred too recently to be strongly considered as synchronous with the warm, dry stage recorded elsewhere. No other systematic trends of forest succession expressive of climatic trends are manifested in the peat profiles from the coastal strip.

PUGET LOWLAND

In the Puget Lowland the replacement of the pioneer lodgepole pine forests by Douglas fir suggests amelioration of the glacial conditions, and undoubtedly there was warming as the influence of the retreating glacier and its causes became more remote. The decline of western white pine upward from the lower levels supports this inference. In most columns western hemlock is recorded as expanding more slowly than Douglas fir which attained its maximum prior to the volcanic activity. This might be interpreted as normal forest succession because hemlock requires better soil conditions than Douglas fir, and is not nearly so aggressive. It is also possible that hemlock was unable to exist upon the sterile soil and under the rigorous conditions left in the wake of the glacier, and required a longer period of time to establish itself on deglaciated terrain. It is significant, however, that in many of the profiles there is a slight decline of hemlock from its maximum, attained below the volcanic ash level, that suggests arrested expansion and even decline prior to the volcanic activity. This trend might also be ascribed to fire, but its synchronous occurrence in so many profiles suggests instead a drier climate as the principal cause. In most of the profiles hemlock begins a second interval of expansion before the volcanic eruption, but the sharpest rise of this species occurs above the ash horizon, and in all profiles it attains its postglacial maximum sometime after this event. The stratum of ash itself was too thin to have had much influence on the vegetation. The consistency of these trends, in view of the fact that hemlock is a moisture-loving species, provides excellent evidence that the influence of the climatic maximum extended to the Puget Lowland. The dry period of the summer probably became longer and was similar in degree to the present conditions in the Willamette Valley where hemlock is largely absent. The abrupt expansion of hemlock above the ash horizon to its postglacial maximum and predominance signifies the return of moister and cooler conditions. The maintenance of these proportions to the present indicates a more or less constant climate since.

In the sedimentary columns located in the drier areas of the cedar-hemlock climax hemlock is recorded in negligible proportions below the ash, and then increases somewhat above this horizon. This suggests that at first the edaphic conditions were unfavorable for hemlock, or it had persisted too far south of the glacier for a rapid expansion into deglaciated areas. Its expansion and partial replacement of Douglas fir was then retarded by the influence of the dry stage. The more recent increase in moisture permitted its limited development, although the conditions have not been so favorable for its predominance over Douglas fir as they have in the Puget Sound region.

WILLAMETTE VALLEY

Climatic interpretations of the pollen profiles from the Willamette Valley must be tempered by the limited records. The replacement of the pioneer forests of lodgepole pine by Douglas fir may be considered as normal forest succession, perhaps partially influenced by the progressively warming and drying climate. The climate reached a degree of desiccation before hemlock had a chance to invade, and in the valley proper summer precipitation has been too light to favor its expansion even after the passing of the climatic maximum. The maximum of oak, attained in the upper part of the Onion Flats and Lake Labish columns, is strong evidence of the warm, dry interval in the Willamette Valley. The absence of the volcanic ash stratum common to the Washington profiles is unfortunate because it fails to provide evidence for the synchroneity of the recorded climatic maximum. The pumice layer apparently represents a later volcanic eruption from a different source. The stratigraphic position of the oak maximum is rather high in the sedimentary columns, but, as already has been shown, the variation in the rate of peat deposition is great. Furthermore, subsidence and deflation of the upper sediments due to drainage and cultivation have truncated the record sufficiently to warrant the inference that the oak maximum was synchronous with the climatic maximum elsewhere recorded.

The persistence of small stands of western yellow pine in the southern part of the Willamette Valley suggests a formerly warmer and drier climate. The pine may be relict of an expansion into the valley at that time. While yellow pine is recorded in some of the profiles, its pollen is too sparse to denote any definite climatic trends.

COLUMBIA BASIN

Perhaps the strongest evidence of the occurrence of the postglacial warm, dry stage in the Pacific Northwest is expressed by the pollen profiles provided by the sedimentary columns located in the Upper Sonoran life zone and Arid Transition area of eastern Washington. The location of five of these profiles within, but near the margin of, the yellow pine forests, has furnished the most significant data. These forests exist under a climate where the available moisture is at a critical minimum, and any slight change in rainfall would result in their advance or retreat and a corresponding adjustment of the adjacent timberless zones. Such a shift apparently happened during the warm, dry stage. The normal expansion of yellow pine to replace the pioneer forests of lodgepole pine and white pine was interrupted by an expansion of grasses, Chenopods, and Composites, which subsided soon after the recorded volcanic activity. Then yellow pine resumed its expansion to a climax status which it has maintained to the present. This rise of xerophytic groups of plants so consistently recorded would seem to be almost irrefutable evidence for the incidence of a warm, dry climate. Its culmination before the volcanic activity and its waning soon after mark its synchroneity with the less strongly expressed climatic maximum of the Puget Sound region. It may be presumed to have been contemporaneous also with the dry stage recorded by oak in the Willamette Valley.

NORTHERN WASHINGTON AND IDAHO

In the montane peat columns of northern Washington and Idaho, remote from the timberless area, the grass maximum attained before the volcanic activity is further evidence for the warm, dry stage. In profiles located in the subclimax white pine forests of northern Idaho white pine was unable to gain predominance over lodgepole pine until after the volcanic activity. It is possible that white pine expansion was retarded by the influence of the warm, dry climate more strongly expressed farther to the south. The occurrence of fire, however, which apparently has been responsible for the persistence of the white pine subclimax, must be considered as having played a major role in forest succession in this region.

NORTHERN GREAT BASIN

The sedimentary columns from the northern Great Basin of south central Oregon, like those in eastern Washington, strongly depict the warm, dry interval. Apparently in this region, also, the precipitation has been at a critical minimum, so that a slight change caused the contraction or expansion of forested areas. In the Lower Klamath Lake profiles the expansion of yellow pine to its maximum at stratigraphic positions that can reasonably be assumed to have been synchronous marks its response to the climatic maximum. A parallel decline of white and whitebark pine is consistent with this recorded trend. Archaeological evidence that early man inhabited an exposed lake bed implies desiccation of the lake and furnishes strong corroborative evidence of warming and drying. The occurrence of the yellow pine maximum near the artifact horizon chronologically correlates these two phases of evidence. The deposition of six to eight feet of fibrous peat above the artifacts on the fossil lake bed indicates reinundation and a return to a cooler and moister climate, as does also an increase in white and whitebark pine contemporaneous with a decline of yellow pine immediately above the artifact horizon.

In the columns from Warner Lake and Chewaucan marshes predominance of grasses, Chenopods, and Composites during the time represented by the middle third of the profiles is believed to portray the warm, dry interval. In the Klamath Marsh section yellow pine has been predominant since the eruption of Mount Mazama, although an appreciable influx of grasses, Chenopods, and Composites records the dry stage.

CASCADE RANGE OF OREGON

Although the deposition of most or all of the peat columns from the Oregon Cascades began prior to the time of maximum dryness, as shown by the relative time of the eruption of Mount Mazama, the dry stage does not seem to be strongly reflected in the pollen profiles. This is partly due to the persistent predominance of lodgepole pine on the pumice mantle and the presence of pollen from two or three life zones interred in the sedimentary columns. In four profiles that lie outside of the pumice mantle, or beyond its influence, the expansion of yellow pine to its maximum in the lower half may have resulted from increased xerism. In the Bend profile, which represents most of postglacial time, the expansion of yellow pine that was interrupted by the pumice fall from Mount Mazama, suggests postglacial drying, but subsequent trends cannot be attributed entirely to climate because of the influence of the pumice in favoring lodgepole pine predominance ever since. The deposition of pumice from later eruptions of other volcanoes has also favored the continued predominance of lodgepole. The most consistent and strongest trend of hemlock occurs above the Mount Mazama pumice horizon and likely expresses moister conditions in more recent times. This evidence is supported by the increase of hemlock in the upper levels of five of the other profiles concurrent with a decrease of yellow pine. In certain profiles a slight rise in white and whitebark pine in the upper half is also suggestive of a return to moister and cooler conditions.

TIME OF THE MOUNT MAZAMA ERUPTION

The records of glaciation on the slopes of Crater Lake mountain and in the rim of the crater, show that Mount Mazama erupted after the maximum of Pleistocene glaciation. The pollen records reveal that it erupted before the time of maximum warmth and dryness, but not until desiccation had reached an advanced degree. The thickness of the sediments overlying Mount Mazama pumice, the stratigraphic position of the interbedded pumice in some of the sections, the proportions and types of sediments overlying and underlying the pumice, and the incidence of

the warm, dry stage as indicated by the profiles in relation to the stratigraphic position of the pumice, all denote that not more than 10,000 years have passed since the climactic eruption of Mount Mazama.

In the Oregon Cascades the degree of yellow pine expansion attained by the time of the eruption suggests that postglacial warming and drying had reached a significant point. The pollen proportions of yellow pine immediately below the pumice layer in those sections containing an interbedded stratum, and the proportions immediately above in those sections southwest of Crater Lake, resting directly upon the pumice, as compared to its proportions in sections located in warmer and drier areas, indicate that the influence of glaciation was long past.

In the Warner Lake section yellow pine attained a proportion of 49 per cent immediately below the pumice stratum, while at Chewaucan Marsh it reached 63 per cent one-half meter below the pumice. In both areas it then declined in favor of grasses, Chenopods, and Composites, disclosing that the climate became too warm and dry for yellow pine even before the eruption of Mount Mazama. If the climatic maximum was reached between 8,000 and 4,000 years ago, then the eruption of Mount Mazama occurred near the end of Period II, or not more than 10,000 years ago, and perhaps even later (table 9).

LATE POSTGLACIAL TIME

The pollen profiles reveal evidence of a moister and cooler climate since the maximum warmth and dryness. In brief this is evidenced by hemlock predominance in the Puget Sound region, decline of oak in the Willamette Valley, decline of grass, Chenopods, and Composites and concurrent expansion of yellow pine in eastern Washington and eastern Oregon, and the slight rise of moisture-loving species in other parts of the Pacific Northwest.

CHRONOLOGY

In making chronological estimates for the major postglacial climatic stages in the Pacific Northwest the best starting points are the present and the volcanic activity recorded by the ash stratum in the Washington sedimentary columns. The volcanic activity responsible for this ash horizon occurred after the culmination of the warm, dry period. An estimate based on data from several sources has been made for the duration of the warm, dry interlude in the Great Basin. This region extends into south central Oregon where several of the sedimentary sections furnishing pollen profiles have been obtained. The proximity of this region, where there is significant evidence for the occurrence of a postglacial warm, dry stage, as well as its length of time, provides a reasonable basis for a correlation of its history with that based upon the pollen profiles from the Pacific Northwest.

SALINITY OF GREAT BASIN LAKES

The present salinity of Owens Lake in California and of Abert and Summer Lakes in south central Oregon is such that it need not have required more than 4,000 years for its development (Gale, 1915; Van Winkle, 1914). Since these lakes have lacked outlets in postglacial time, their present low degree of salinity indicates that their pluvial antecedents dried up and the accumulated salts thereof were removed by wind or were buried before the modern lakes came into existence. Antevs (1938, 1946) interprets this to mean that these lakes, as well as many others in the Great Basin, dried up during the dry Middle Postpluvial (Postglacial) and then were reborn with the advent of increased moisture in later postglacial time. He sets the duration of this warm, dry maximum from about 8,000 to 4,000 years ago.

MODERN GLACIERS OF WESTERN MOUNTAINS

Further evidence of the warm, dry maximum during this time is revealed by the history of modern glaciers in the western mountains. All of the fifty modern cirque glaciers in the Sierra Nevada, almost all the glaciers in the Rocky Mountains within the United States, and all of the lesser glaciers of the Cascade Range and the Olympic Mountains may represent a new generation of glaciers that came into existence in relatively recent time, probably about 4,000 years ago (Matthes, 1939, 1942). Before this there had been almost complete disappearance of permanent ice in the mountains, so as to suggest a long, warm interval.

EROSION IN THE SOUTHWEST

Arid conditions during the middle Postpluvial are also shown by arroyo cutting and wind erosion in Arizona, New Mexico, and western Texas, and by an exceptionally low level of Utah Lake (Antevs, 1941a; Hansen, 1933).

CORRELATION OF PUMICE STRATA

The eruption of Mount Mazama during postglacial time (Williams, 1942) also apparently took place before the culmination of the warm, dry stage. This is shown by three layers of Crater Lake pumice that are interbedded in lacustrine sediments in the former bed of pluvial Lake Chewaucan and Winter Lake, predecessors of Summer Lake in south central Oregon (Allison, 1945). A formerly deep lake was still in existence several tens of feet above modern Summer Lake level at the time of the final major eruption. That it persisted for some time afterward is disclosed by additional lake sediments overlying the Crater Lake pumice, including a layer of pumice from Newberry Crater to the north-northwest. It is believed that Winter Lake dried up during the warm, dry

Middle Postglacial, as did many other lakes of the Great Basin. Subsequently, Summer Lake was reestablished in the lowest part of the basin. If the fluctuations of Chewaucan, Winter, and Summer lakes were correlative with those of other lakes in the Great Basin, the final eruption of both Mount Mazama and Newberry Crater occurred before the culmination of the warm, dry stage in the Great Basin. The desiccation of Lower Klamath Lake, the recorded occupancy of the exposed lake bed by early man, and the warm, dry maximum revealed by the accompanying pollen profiles are entirely consistent with the other evidence.

VOLCANIC ASH STRATA IN WASHINGTON PROFILES

The pollen profiles from Washington reveal strong evidence of a warm, dry stage that reached its maximum sometime before the recorded volcanic activity and waned soon afterward. This climatic maximum is reflected by the trends not only of several different species but also in areas of different present climate. Accepting the dates for the warm, dry interval as from 8,000 to 4,000 years ago, the volcanic eruption in Washington must have occurred about 6,000 years ago, or near the middle of the warm, dry middle postglacial stage. Because of the greater marine influence in the Puget Sound region, the duration of the warm, dry interval was probably somewhat shorter than in the Willamette Valley and Columbia Basin (table 9)

CORRELATION WITH EUROPE AND EASTERN NORTH AMERICA

There is little or no evidence of alternating dry and moist stages during middle postglacial time, such as are set forth by the Blytt-Sernander hypothesis for northern Europe, and as have also been interpreted from eastern North American pollen profiles by Sears and Deevey. The first stage of warmth and dryness apparently was not replaced by a moister phase equivalent to the Atlantic, to be succeeded by a final interval of warmth and dryness corresponding to the sub-boreal or the classical xerothermic stage (Sears, 1942). Instead there was a single middle postglacial stage of maximum warmth and dryness, followed by cooling and an increase in moisture to a degree which has remained more or less constant to the present. If there was an increase in moisture corresponding to that of the Atlantic stage, one might expect it to be magnified in a region where a marine climate existed, unless the climate was already wet enough and moisture was not a limiting factor. There seems to be no evidence in the pollen profiles discernible to the author for such a trend.

The European sub-boreal or second stage of warmth and dryness persisted until about 2,600 years ago (table 8). The stratigraphic position of the Washing-

ton volcanic ash, well below the middle of most profiles, hardly permits the persistence of the warm,
dry stage in the Pacific Northwest to less than 4,000
years ago. Thus, the interpretation of Pacific Northwest pollen profiles bears out a climatic sequence
similar to that of von Post and Antevs; namely, about
10,000 years of progressive warming and drying, a
period of higher temperature than now prevails for
another 4,000 years, followed by cooling and an increase in moisture to a degree that has persisted to the
present, or for about 4,000 years. The final cycle of
cooling was more rapid than the development of the
stage of maximum warmth. The last two stages were
approximately synchronous with the second and third
stage of von Post, but the initial stage of warming and
drying was considerably longer owing to the earlier
retreat of the ice from Washington than from northern
Europe. The duration of the warm, dry interval in
the Pacific Northwest was perhaps shorter than the
combined European boreal, Atlantic, and sub-boreal
stages. Its shortest régime seems to have been in the
Puget Sound region where a pronounced return to
moister conditions was already under way by the time
of the recorded volcanic activity. On the Pacific
Coast the influence of the marine climate apparently
was never overcome during postglacial time.

RECENT CLIMATIC TRENDS

The warm, dry middle Postglacial ended about
4,000 years ago in the West, and cooler and moister
conditions have since prevailed almost to the present.
This is indicated by the accumulation of water in the
desert lake basins, the rebirth of glaciers in the high
mountains, and the expansion of moisture-loving
forest trees in the Pacific Northwest. In eastern
North America an increase of spruce and fir in more
recent time is interpreted as cooling of the climate
almost to the present. In England an increase in
birch and beech and a correlative decline of oak, alder,
and basswood since the warm, dry middle postglacial
period is interpreted to mean the return of moister
and cooler conditions (Godwin, 1940). In Poland an
expansion of spruce during later postglacial time is
ascribed to cooling, which is also imputed to central
Europe in general (Szafer, 1935).

RECENT GLACIAL MOVEMENTS

Whereas the evidence suggests cooling and more
moisture since about 4,000 years ago, there is evidence
from various sources for warmer and drier conditions
during the last few centuries. Such climatic variations may be of short duration and may not represent
the beginning of a long range cycle equivalent to the
major trends of postglacial time. In the high western
mountains the glaciers attained their maximum expansion of the past 10,000 years in about A.D. 1850
(Matthes, 1942). In Sweden the precipitation had

its postglacial maximum in the centuries just before
Christ (Granlund, 1932), while the Scandinavian
glaciers reached their greatest postglacial size during
the age about A.D. 1740 to 1825. Great Basin lakes
reached unusually high levels about 1870 (Woolley,
1924; Antevs, 1938). Since about 1850 the temperature has been rising in the United States, the rate
being especially marked since about 1900 (Kincer,
1940). In Alaska there has been a general recession
of glaciers in the last two centuries (Cooper, 1942),
while the same is true for the Pacific Northwest.
Freeman (1941) found that Lyman Glacier in northern Washington has retreated at an average rate of
60 feet per year for the past eleven years, and that the
average rate of retreat since the 1890's has been about
40 feet per year. Since the beginning of the century
some of the principal glaciers in Glacier National
Park, Montana, have been reduced 40 to 75 per cent
in area and even more in volume (Dyson, 1941). In
interpreting the meaning of present glacial retreat it
would seem that the size of the glacier would be important. The present retreat of a main trunk glacier
may in part reflect the conditions of a century or more
ago, and present changes in precipitation may not be
evidenced by these glaciers for another century. The
oscillations of smaller glaciers, however, probably
represent more recently changing conditions, perhaps
a few years previous.

TREE-RING RECORDS

The tree-ring records for eastern and south central
Oregon indicate that during the past 650 and 475 years
respectively there has been no general trend toward
drier or wetter climate (Keen, 1937; Antevs, 1938a).
The record does show, however, that a critical subnormal growth period existed from 1917 or 1918 to
1935 or 1936. As compared with other drouth
periods, this one was the most severe and critical
during the entire time covered by the above records.
In eastern Oregon other periods have exceeded it in
duration of subnormal growth, but none has ap-
proached it in severity. Growth in 1931, the poorest
year, was 68 per cent below normal. The decade
from 1670 to 1680 was evidently the wettest within
the life span of present living pine forests (Keen,
1937). In south central Oregon the drouth of 1918–
1935 was both the longest and the most severe of
record, while that of 1840–1850 was second in degree
of severity. The 67-year period from 1851 to 1917
may have been moister than any similar period since
1455 (Antevs, 1938a). In analyzing the California
precipitation record from 1850 to 1933, Gray (1934)
found a downward trend amounting to about 8 inches
in 80 years. For the same period the tree-ring record
shows a downward trend correlative with the precipitation record.

PLANT SUCCESSION ON BOGS

Recent succession of species of *Sphagnum* in English bogs, with the almost complete elimination of *Sphagnum imbricatum* at present, may indicate increased dryness and continentality (Godwin, 1940). This same succession has been noted in Holland and northwest Germany.

PACIFIC NORTHWEST POLLEN PROFILES

In support of the foregoing evidence for warming and drying during the very recent time, the pollen profiles of the Pacific Northwest can offer little. In some profiles the recorded trends seem to provide a basis for such an interpretation, while in others a more humid trend is depicted. Most of the fluctuations recorded immediately below and at the top are of too small magnitude to warrant an interpretation of vegetation change. That there has not been continual increase in moisture to the present is suggested by the stability of moisture-loving species after they had attained their maxima succeeding the warm, dry middle postglacial stage. Other factors which certainly have influenced the uppermost pollen proportions are lumbering, fire, and cultivation since the advent of white man. It must be admitted that in some profiles the trend may reflect the changes attendant to the removal of timber cover, while in others it is quite opposed to what one would expect. In the Douglas fir region, where this species has provided the bulk of the commercial timber since lumbering operations began in the Pacific Northeast, some profiles record a final increase and others a decline of this species.

Whereas there is strong evidence that, since the warm, dry middle postglacial stage maximum humidity prevailed over the northern hemisphere several centuries ago, there does not seem to be sufficient proof of sustained warming and drying during the last few centuries that may be expected to continue into the future. The span of time is in itself too short to warrant any such interpretation, and any present-day trends may be only parts of short cycles of minor climatic oscillations superimposed on a more general, though as yet obscure, major trend.

EARLY MAN IN THE PACIFIC NORTHWEST

Correlation of pollen-analytical chronology with European prehistory has reached a significant stage. Godwin (1940) has linked postglacial forest history with human prehistory and correlated pollen profiles from the British fenlands with human culture from the Mesolithic up through the Romano-British. The former dates back to at least the pre-boreal which was characterized by forests of birch and pine. Correlations with De Geer's varved clay chronology suggests that it goes back at least 10,000 years.

In North America the record of early man is less distinct, and until 15 or 20 years ago the possibility of his existence was minimized. Since then, however, evidence has been discovered that points to man's presence in America probably well over 10,000 years ago. These discoveries show without a doubt the association of man with an extinct fauna typical of the late Pleistocene or early Postglacial, extending over a vast area from far north in the High Plains, south nearly to the Gulf of Mexico, westward to the Colorado River drainage, and northward to the northern limits of the Great Basin in southern Oregon (Howard, 1935; Roberts, 1935, 1936; Sayles and Antevs, 1941; Bryan and Ray, 1940; Cressman, 1942). Since about 18,000 years ago there has been an ice-free passage into the United States by way of the Yukon and Mackenzie River regions of Alaska and Canada and a corridor between the shrinking Cordilleran and Keewatin ice sheets at the eastern foot of the Canadian Rockies (Antevs, 1946). When migrating man reached the Missouri River in western Montana, he could have followed this valley to the mountain passes opening into the Snake River drainage and continued his migration into the Great Basin. He first reached the Southwest during the last part of the Pluvial epoch over 10,000 years ago (Antevs, 1935). That he must have arrived in the northern Great Basin in south central Oregon as early is shown by the evidence for his existence before the eruption of Mount Mazama about 10,000 years ago, prior to the warm, dry middle postglacial stage which began about 8,000 years ago. If he went from north to south, it could reasonably be expected that he was in the northern Great Basin before he reached the Southwest.

The existence of early man in the northern Great Basin was apparently closely related to the history of its lakes. These lakes, born during the Pleistocene, fluctuated in response to glacial cycles. They attained their highest levels during the Iowan-Tahoe glacial stage. During the following interglacial they receded, and then assumed another high level, but lower than the first, during the late Wisconsin-Tioga glacial stage. There are seven of these lakes in south central Oregon, occupying a series of basins extending eastward from the Cascade Range to the extreme southeastern portion of the state. With the exception of Klamath Lake on the western border of the province, they have interior drainage and, from the position of their shorelines, have undoubtedly fluctuated in response to the changing climate during the Postglacial as well as during the Pleistocene. Long before man reached the northern Great Basin early Wisconsin lakes had cut the highest terraces and caves in the cliffs along these shorelines. When man arrived on the scene the late Wisconsin lakes had probably receded owing to postglacial warming and drying. There was still sufficient fresh water in these lakes to supply the needs of early man and the animals upon which he depended, however, and he occupied the caves, As the lakes evaporated during the warm, dry middle Postglacial, he waś forced to migrate to be near a constant water supply. He has left a record of this migration from east of the Steens Mountains to the Cascade Range to the west (Cressman, 1941). Evidence suggests contemporaneous occupation of the Willamette Valley, but this is not at present convincing (Cressman and Laughlin, 1941). The record of man during the dry interval is less definite than before and after, but that he was able to survive goes without saying.

WIKIUP DAM SITE

Obsidian knives were found at the Wikiup Dam Site on the Deschutes River about thirty miles south of Bend, Oregon (Cressman, 1937). The artifacts were imbedded in a very compact bed of sand at some distance underneath pumice from Mount Mazama, indicating that the Paleo-Indians lived there well before the pumice fall, and therefore some time during the early Postpluvial.

PAISLEY CAVES

Mount Mazama pumice overlies human artifacts in caves near Paisley, Oregon, in Summer Lake Valley (Cressman, Krieger, and Williams, 1940; Cressman, 1942). These caves were excavated by waves when the level of Pluvial (glacial) Lake Chewaucan (Allison, 1945) was at its maximum height in Bonneville (Tahoe-Iowan) time. The earliest occupation of the caves is probably contemporaneous with the emplacement of the obsidian knives at the Wikiup Dam Site. The clean character of the pumice in the Paisley caves shows that it was not blown in during the warm, dry, middle Postpluvial but that it either was deposited directly from the air by the clouds of pumice ejected at the time of the eruption, or was blown in shortly thereafter from drift pumice on snow or from gravity deposits at the mouth of the cave. The antiquity of the earliest occupation of these caves is shown by the association of the artifacts with the bones of extinct fauna: horse, camel, and other genera indicative of a moister climate than now prevails (Cressman, 1942).

FORT ROCK CAVE

In the Fort Rock Cave, located in the Fort Rock Basin north of the Summer Lake Basin, occupational debris is overlain by pumice from Newberry Crater, located about 20 miles to the north (Cressman, Williams, and Krieger, 1940). Although the eruption of Newberry Crater was later than that of Mount Mazama, it occurred just prior to the middle postglacial dry stage, as shown by the relative stratigraphic position of the pumice in the Pluvial Winter Lake sediments (Allison, 1945).

LOWER KLAMATH LAKE

The correlation of the pollen record with the presence of early man in the Klamath Basin has already been discussed. Since Lower Klamath Lake existed in symbiotic relationship to Upper Klamath Lake, it was probably affected by the arid middle post-Pleistocene somewhat later than the other basin lakes. At any rate there was water in the Upper Klamath Lake and its outlet. The evidence from Lower Klamath Lake indicates that man was present at the same time as the horse, camel, and some form of elephant. The Klamath Lake area was thus a favored spot because of the persistence of water, and it is possible that the Pleistocene fauna lingered on somewhat longer in this more favorable area than farther to the east. Perhaps some of the occupants of the eastern area moved westward, following desiccation of the lakes, to join with the occupants of the Klamath country. Eventually, between 8,000 and 4,000 years ago, the drouth became severe enough to dry up Lower Klamath Lake and make that uninhabitable. At the beginning of the late Postpluvial with the gradual increase of precipitation water was again found in the Lower Klamath Lake, and the Indians began again to camp along the shoreline as is shown by the stratigraphic position of the artifacts in the Lairds Bay locality, as previously discussed in connection with the pollen profiles in that area.

EASTERN NORTH AMERICA

The only other known correlation of pollen analysis with archeology in America has been made in New England. A fishweir was excavated in Boston, about 15 feet below the present sea level. Pollen analysis of the sediments surrounding the fishweir and its correlation with that of a nearby peat profile of greater age suggest that it was built near the end of a dry period similar in character and position to the sub-boreal of European chronology (Knox, 1942). The fishweir was in use, according to tentative figures, about 300 years, and was abandoned near the end of the dry period, tentatively dated at 1200 B.C.

REFERENCES

ALLISON, I. S. 1935. Glacial erratics in Willamette Valley. *Geol. Soc. Amer., Bull.* **46**: 615–632.
—— 1936. Pleistocene alluvial stages in northwestern Oregon. *Science* **83**: 441–443.
—— 1945. Pumice beds at Summer Lake, Oregon. *Geol. Soc. Amer., Bull.* **56**: 789–808.
ANDERSSON, GUNNAR. 1892. Om de växgeografiska och växtpaleontologiska stöden för antagandet af klimatväxlinger under kvärtartiden. *Geolog. Fören. Stockh. Förhandl.* **14**: 509–538; *Bot. Centr.* **55**–56; 48–51.
ANTEVS, ERNST. 1922. Recession of the last ice sheet in New England. *Amer. Geogr. Soc. Res. Ser.* No. **11**: 1–120.
—— 1925. On the Pleistocene history of the Great Basin. *Carnegie Inst. Wash. Publ.* No. **352**: 51–144.
—— 1925a. Swedish late-Quaternary geochronologies. *Geogr. Rev.* **15**: 280–284.
—— 1928. The last glaciation. *Amer. Geogr. Soc. Res. Ser.* No. **17**: 1–292.
—— 1929. Maps of the Pleistocene glaciation. *Geol. Soc. Amer., Bull.* **40**: 631–720.
—— 1931. Late glacial correlations and ice recession in Manitoba. *Geol. Surv. Canada Mem.* **169**: 1–76.
—— 1933. Correlations of late Quaternary chronologies. *Rept. 16th International Congress*, 1–4.
—— 1935. The spread of aboriginal man to North America. *Geogr. Rev.* **25**: 302.
—— 1938. Postpluvial climatic variations in the Southwest. *Bull. Amer. Meteor. Soc.* **19**: 190–193.
—— 1938a. Rainfall and tree growth in the Great Basin. *Carnegie Inst. Wash. Publ.* 469; *Amer. Geogr. Soc., Spec. Publ.* 21.
—— 1939. Late Quaternary upwarpings of northeastern North America. *Amer. Jour. Geol.* **48**: 707–720.
—— 1940. Age of artifacts below peat beds in Lower Klamath Lake, California. *Carnegie Inst. Wash. Yearbook* No. 39; 307–309.
—— 1941. Age of the Cochise Culture stages. *Medallion Papers* No. 29; 31–55. Globe, Arizona.
—— 1941a. Climatic variations in the Southwest during the past 75,000 years. *Pan-Amer. Geol.* **76**: 73–75.
—— 1945. Correlation of Wisconsin glacial maxima. *Amer. Jour. Sci.* **243**-A: 1–39. (Daly Volume.)
—— 1946. The Great Basin, with emphasis on glacial and postglacial times. III. Climatic changes and pre-white man. MS.
ASSARSSON, G., and E. GRANLUND. 1924. En metod for pollenanalys av minerogena jordarter. *Geol. Foren. Forhandl.* **46**.
AUER, V. 1927. Stratigraphic and morphological investigations of peat bogs in southeastern Canada. *Com. Inst. Quaetonum Foresta, Finlandiae*, Edita 12.
—— 1930. Peat bogs in southern Canada. *Dept. Mines, Geol. Surv., Canada*, Memoir 162.
BAILEY, V. 1936. The mammals and life zones of Oregon. *North Amer. Fauna* **55**: 1–146. U. S. Dept. Agric., Washington, D. C.
BAKER, F. S. 1936. Theory and practice of silviculture. N. Y., McGraw.
BARKLEY, F. 1934. The statistical theory of pollen analysis. *Ecology* **15**: 283–289.
BERTSCH, K. 1931. Paläobotanische Monographie des Federseerieds, *Bibliotheca Bot.* **26**: Hefte 103: 1–127.
BLACKWELDER, E. 1931. Pleistocene glaciation in the Sierra Nevada and Basin Ranges. *Geol. Soc. Amer., Bull.* **42**: 865–922.

BLYTT, A. 1881. Die Theorie der wechselnden continentalen und insularen Klimate. *Bot. Jahrb.* **2**: 1–50, 177–184. *Bot. Centr.* **7**: 299–308.
BOWMAN, P. W. 1931. Study of a peat bog near the Matamek River, Quebec, Canada, by the method of pollen analysis. *Ecology* **12**: 694–708.
BRETZ, J. H. 1913. Glaciation of the Puget Sound region. *Wash. Geol. Surv. Bull.* No. **8**: 1–244.
—— 1923. Glacial drainage of the Columbia Plateau. *Geol. Soc. Amer., Bull.* **34**: 573–609.
BRYAN, K., and L. L. RAY. 1940. Geological antiquity of the Lindenmeier site in Colorado. *Smithsonian Misc. Coll.* 99.
CAIN, S. A. 1940. The identification of species in fossil pollen of Pinus by size-frequency determinations. *Amer. Jour. Bot.* **27**: 301–308.
CAIN, S. A., and L. G. CAIN. 1944. Size-frequency studies of *Pinus palustris* pollen. *Ecology* **25**: 229–233.
CARROLL, GLADYS. 1943. The use of Bryophytic polsters and mats in the study of recent pollen deposition. *Amer. Jour. Bot.* **30**: 361–366.
COOPER, W. S. 1942. Vegetation of the Prince Williams Sound region, Alaska; with a brief excursion into post-Pleistocene climatic history. *Ecol. Monogr.* **12**: 1–22.
CRESSMAN, L. S. 1937. The Wikiup damsite no. 1 knives. *Amer. Antiquity* **3**: 53–67.
—— 1940. Studies on early man in south central Oregon. *Carnegie Inst. Wash. Year Book* No. 39: 300–306.
—— 1941. Early man in the northern Great Basin of south central Oregon. *6th Pacific Sci. Congr., Proc.* **4**: 169–175.
—— 1942. Archaeological researches in the Northern Great Basin. *Carnegie Inst. Wash. Publ.* No. 538: 1–155.
CRESSMAN, L. S., H. WILLIAMS, and A. D. KRIEGER. 1940. Early man in Oregon. *Univ. Monogr., Studies in Anthropology* No. 3.
CRESSMAN, L. S., and W. S. LAUGHLIN. 1941. A probable association of man and the mammoth in the Willamette Valley, Oregon. *Amer. Antiquity* **6**: 339–343.
CROLL, JAMES. 1875. Climate and time in their geological relations. London, Daldy, Isbister.
DACHNOWSKI, A. P. 1926. Profiles of peat deposits in New England. *Ecology* **7**: 120–135.
DACHNOWSKI-STOKES, A.P. 1936. Peat lands in the Pacific coast states in relation to land and water resources. *U. S. Dept. Agric., Misc. Publ.* 248.
—— 1941. Peat resources in Alaska. *U. S. Dept. Agric. Tech. Bull.* No. 769.
DALY, R. A. 1929. Swinging sea level of the ice age. *Geol. Soc. Amer., Bull.* **40**: 721–734.
—— 1934. The changing world of the ice age. Yale Univ. Press.
DAUBENMIRE, R. F. 1942. An ecological study of the vegetation of southeastern Washington and adjacent Idaho. *Ecol. Monogr.* **12**: 53–79.
DEEVEY, E. S. 1939. Studies on Connecticut lake sediments, I. A postglacial climatic chronology for southern New England. *Amer. Jour. Sci.* **237**: 691–724.
—— 1942. Studies on Connecticut lake sediments, III. The biostratonomy of Linsley pond. *Amer. Jour. Sci.* **240**: 233–264; 313–338.
—— 1943. Additional pollen analyses from southern New England. *Amer. Jour. Sci.* **241**: 717–752.
—— 1944. Pollen analysis and history. *Amer. Scientist* **32**: 39–53.
DE GEER, G. 1910. A geochronology of the last 12,000 years. *Congr. Geol. Int.* 11 *Compte rendu* **1**: 241–253.

—— 1940. Geochronologia suecica. Principles. *Svenska Vetenscapsakad. Handl.* 18: No. 6. Stockholm.

DICE, L. R. 1943. The biotic provinces of North America. Univ. of Mich.

DILLER, J. S. 1896. A geological reconnaissance in northwestern Oregon. *U. S. Geol. Surv., Ann. Rept.* 17 (1): 480–483.

—— 1915. The relief of our Pacific Coast. *Science* 41: 48–57.

DRAPER, P. 1928. A demonstration of the technique of pollen analysis. *Proc. Oklahoma Acad. Sci.* 8: 63–64.

DYSON, J. L. 1941. Recent glacier recession in Glacier National Park, *Montana. Jour. Geol.* 49: 815–825.

ERDTMAN, G. 1931. Pollen statistics: a new research method in paleoecology. *Science* 73: 399–401.

—— 1936. New methods in pollen analysis. *Svensk. Bot. Tidsskrift* 30: 154–164.

—— 1943. An introduction to pollen analysis. Waltham, Mass., Dawson.

FENNEMAN, N. M. 1931. Physiography of western United States. N. Y., McGraw.

FIRBAS, F. 1934. Über die Bestimmung der Walddichte und der vegetation waldloser Gebeite mit Hilfe der pollenanalys. *Planta* 22: 109–145.

FLINT, R. F. 1937. Pleistocene drift border in eastern Washington. *Geol. Soc. Amer., Bull.* 48: 203–233.

FREEMAN, O. W. 1941. Recession of Lyman Glacier, Washington. *Jour. Geol.* 49: 764–771.

FRÜH, J. 1885. Kritische Beitrage zue Kenntnis des Torfes. *Jahrb. K. K. Reichsanstalt* 35.

FULLER, G. D. 1935. Postglacial vegetation of the Lake Michigan region. *Ecology* 16: 473–487.

GALE, H. S. 1915. Salines in the Owens, Searles, and Panamint basins, southeastern California. *U. S. Geol. Surv. Bull.* 580: 251–323.

GEISLER, F. 1935. A new method for separation of fossil pollen. *Butler Univ. Bot. Studies* 3: 141–146.

GODWIN, H. 1940. Pollen analysis and forest history of England and Wales. *New Phytologist* 39: 370–400.

GRANLUND, ERIK. 1932. De svenska hogmossarnas geologi Sveriges Geol. Undersokning, Ser. C, No, 373, Arsbok 26. Stockholm.

—— 1936. *In* Magnusson, Nils and Erik Granlund. Sveriges geologi. P. A. Norstedt and Söner Stockholm.

GRAY, L. G. 1934. Long-period fluctuations of some meteorological elements in relation to California forest-fire problems. *Mo. Weath. Rev.* 62: 231–235.

HANSEN, G. H. 1933. An interpretation of past climatic cycles from observations made of Utah Lake sediments. *Utah Acad. Sci., Arts, and Letters* 11: 162–163.

HANSEN, H. P.[6] 1933. The Tamarack bogs of the Driftless Area of Wisconsin. *Bull. Pub. Mus. Milwaukee* 7: 231–304.

—— 1937. Pollen analysis of two Wisconsin bogs of different age. *Ecology* 18: 136–148.

—— 1938. Postglacial forest succession and climate in the Puget Sound region. *Ecology* 19: 528–542.

—— 1939a. Pollen analysis of a bog in northern Idaho. *Amer. Jour. Bot.* 26: 225–228.

—— 1939b. Pollen analysis of a bog near Spokane, Washington. *Bull. Torrey Bot. Club* 66: 215–220.

—— 1939c. Postglacial vegetation of the Driftless Area of Wisconsin. *Amer. Midl. Nat.* 21: 752–762.

—— 1939d. Paleoecology of a central Washington bog. *Ecology* 20: 563–568.

—— 1940a. Paleoecology of two peat bogs in southwestern British Columbia. *Amer. Jour. Bot.* 27: 144–149.

—— 1940b. Paleoecology of a montane peat deposit at Bonaparte Lake Washington. *Northwest Science* 14: 60–69.

—— 1941a. Paleoecology of a bog in the spruce-hemlock climax of the Olympic Peninsula. *Amer. Midl. Nat.* 25: 290–297.

—— 1941b. Further pollen studies of post-Pleistocene bogs in the Puget Lowland of Washington. *Bull. Torrey Bot. Club* 68: 133–148.

—— 1941c. Paleoecology of a peat deposit in west central Oregon. *Amer. Jour. Bot.* 28: 206–212.

—— 1941d. Paleoecology of two peat deposits on the Oregon Coast. *Oregon State Monogr.; Studies in Botany* 3: 1–31.

—— 1941e. A pollen study of post-Pleistocene lake sediments in the Upper Sonoran life zone of Washington. *Amer. Jour. Sci.* 239: 503–522.

—— 1941f. Paleoecology of a montane peat deposit near Lake Wenatchee, Washington. *Northwest Science* 15: 53–65.

—— 1942a. The influence of volcanic eruptions upon post-Pleistocene forest succession in central Oregon. *Amer. Jour. Bot.* 29: 214–219.

—— 1942b. A pollen study of lake sediments in the lower Willamette Valley of western Oregon. *Bull. Torrey Bot. Club* 69: 262–280.

—— 1942c. Post-Mount Mazama forest succession on the east slope of the central Cascades of Oregon. *Amer. Midl. Nat.* 27: 523–534.

—— 1942d. A pollen study of peat profiles from Lower Klamath Lake of Oregon and California. In: Archaeological researches in the Northern Great Basin. *Carnegie Inst. Wash. Publ.* No. 538: 1–155. L. S. Cressman.

—— 1942e. A pollen study of a montane peat deposit near Mount Adams, Washington. *Lloydia* 5: 305–313.

—— 1943a. A pollen study of a subalpine bog in the Blue Mountains of northeastern Oregon. *Ecology* 24: 70–78.

—— 1943b. A pollen study of two bogs on Orcas Island, of the San Juan Islands, Washington. *Bull. Torrey Bot. Club* 70: 236–243.

—— 1943c. Paleoecology of two sand dune bogs on the southern Oregon Coast. *Amer. Jour. Bot.* 30: 335–340.

—— 1943d. Paleoecology of a peat deposit in east central Washington. *Northwest Science* 17: 35–40.

—— 1943e. Post-Pleistocene forest succession in northern Idaho. *Amer. Midl. Nat.* 30: 796–803.

—— 1944. Further pollen studies of peat bogs on the Pacific Coast of Oregon and Washington. *Bull. Torrey Bot. Club* 71: 627–636.

—— 1944. Postglacial vegetation of eastern Washington. *Northwest Science* 18: 79–87.

—— 1946. Postglacial forest succession in the Oregon Cascades. *Amer. Jour. Sci.* 244: 710–734.

—— 1946. Early man in Oregon: Pollen analysis and postglacial climate and chronology. *Scientific Monthly* 62: 52–62.

HANSEN, H. P., and I. S. ALLISON. 1942. A pollen study of a fossil peat deposit on the Oregon Coast. *Northwest Science* 16: 86–92.

HANZLIK, E. J. 1932. Type succession in the Olympic Mountains. *Jour. Forestry* 30: 91–93.

HARLOW, W. M., and E. S. HARRAR. 1941. Textbook of dendrology. N. Y. McGraw.

HOBBS, W. H. 1943. Discovery in eastern Washington of a new lobe of the Pleistocene continental glacier. *Science* 98: 227–230.

HOFMANN, J. V. 1924. Natural regeneration of Douglas fir in the Pacific Northwest. *U. S. Dept. Agric. Bull.* 1200: 1–63.

HOLLICK, A. 1931. Plant remains from a Pleistocene lake deposit in the upper Connecticut River Valley. *Brittonia* 1: 35–55.

HORMANN, H. 1929. Die pollenanalytische Unterscheidung von *Pinus montana, P. silvestris,* and *P. ecmbra. Oester. Bot. Zeit.* 78: 215–228.

HOWARD, E. B. 1935. Evidence of early man in North America based on geological and archaeological work in New Mexico. *Smithsonian Inst. Misc. Coll.* 94: no. 4.

HUBERMAN, M. A. 1935. The role of eastern white pine in forest succession in northern Idaho. *Ecology* 16: 137–152.

[6] All papers by the author on pollen analysis of bogs of the Pacific Northwest are listed here, whether cited in the text or not.

ISAAC, L. A. 1943. Reproductive habits of Douglas fir. Washington, D. C., U. S. Forest Service; Charles Lathrop Pack Forestry Foundation.

JAESCHKE, J. 1935. Zur Frage der Artdiagnose der *Pinus silvestris*, *P. montana*, und *P. cembra* durch variationstatistische Pollenmessungen. *Beih. Bot. Cent.* **52B**: 622–633.

JONES, G. N. 1936. A botanical survey of the Olympic Peninsula. *Univ. Wash. Publ.* **6**: 1–286.

—— 1938. The flowering plants and ferns of Mount Rainier. *Univ. Wash. Publ. Biol.* **7**: 1–192.

JUDD, C. S. 1915. Douglas fir and fire. *Proc. Soc. Amer. Foresters* **10**: 186–191.

KAY, C. F. 1931. Classification and duration of the Pleistocene period. *Geol. Soc. Amer., Bull.* **42**: 425–466.

KEEN, F. P. 1939. Insect enemies of western forests. *U. S. Dept. Agric. Misc. Publ.* **273**: 1–210.

—— 1937. Climatic cycles in eastern Oregon as indicated by tree rings. *Mo. Weath. Rev.* **65**: 175–188.

KINCER, J. B. 1940. Relation of recent glacier recession to prevailing temperatures. *Mo. Weath. Rev.* **68**: 158–160.

KNOX, A. S. 1942. The pollen analysis of silt and tentative dating of the deposits. *In* JOHNSON, F. *et al.* The Boyleston Street Fishweir. *Robert S. Peabody Foundation for Archaeology Papers* **2**: 105–129.

LARSEN, J. A. 1929. Fires and forest succession in the Bitterroot Mountains of northern Idaho. *Ecology* **10**: 67–76.

LAWRENCE, D. B. 1938. Trees on the march. *Mazama Annual.* Portland.

—— 1941. The floating island lava flow of Mt. St. Helens. *Mazama* **23**: 56–60.

LESQUEREUX, L. 1885. On the vegetable origin of coal. *Pa. Geol. Surv. Ann. Rept.*, 95–124.

LEVERETT, F., and F. W. SARDESON. 1932. Quaternary geology of Minnesota and parts of adjacent states. *U. S. Geol. Surv. Prof. Paper* **161**: 1–149.

LIDÉN, RAGNAR. 1938. Den senkvartära strandförskjutningens förlopp och kronologi i Ångermanland. *Geol. Fören. i Stockholm Förhandl.* **60**: 397–404.

LIVINGSTON, B. E., and F. SHREVE. 1921. The distribution of vegetation in the United States, as related to climatic conditions. *Carnegie Inst. Wash. Publ.* No. 284.

LINDEMAN, R. L. 1941. The developmental history of Cedar Creek bog, Minnesota. *Amer. Midl. Nat.* **25**: 101–112.

LÜDI, W. 1937. Die pollensedimentation in Davoserhochtale. *Ber. Geobot. Forschungsinstitut Rübel, Zürich 1936*: 107–126.

MacCLINTOCK, PAUL, and E. T. APFEL. 1944. Correlation of the drifts of the Salamanca re-entrant, New York. *Geol. Soc. Amer. Bull.*, **55**: 1143–1164.

MACKIN, J. H. 1941. Glacial geology of the Snoqualmie-Cedar area, Washington. *Jour. Geol.* **49**: 449–481.

MATTHES, F. E. 1939. Report of the Committee on Glaciers. *Trans. Amer. Geophysical Union.* **4**: 518–523.

—— 1942. Glaciers. *In* Hydrology. N. Y., McGraw.

MERRIAM, C. H. 1894. Life zones and crop zones of the United States. *U. S. Dept. Agric., Biol. Surv. Bull.* **10**: 1–79.

MOVIUS, H. L. 1942. The Irish Stone Age. Cambridge Univ. Press.

MUNGER, T. T. 1940. The cycle from Douglas fir to hemlock. *Ecology* **21**: 451–459.

MUNNS, E. N. 1938. The distribution of important forest trees of the United States. *U. S. Dept. Agric. Misc. Bull.* **287**: 1–176.

NIKIFOROFF, C. C. 1937. The inversion of the great soil zones in western Washington. *Geog. Rev.* **27**: 200–213.

OSBORN, H. F., and C. A. REEDS. 1922. Old and new standards of Pleistocene division in relation to the prehistory of man in Europe. *Geol. Soc. Amer., Bull.* **33**: 411–490.

OSVALD, H. 1936. Stratigraphy and pollen flora of some bogs of the North Pacific Coast of America. *Ber. Schweiz. Bot. Ges.* **46**.

PAGE, B. M. 1939. Multiple glaciation in the Leavenworth area, Washington. *Jour. Geol.* **47**: 785–816.

PECK, M. E. 1941. A manual of the higher plants of Oregon. Portland, Binfords.

PIPER, C. V. 1906. Flora of the state of Washington. *Contr. U. S. Nat. Herb.* **11**: 1–637.

POTZGER, J. E. 1942. Pollen spectra from four bogs on the Gillen Nature Reserve, along the Michigan-Wisconsin state line. *Amer. Midl. Nat.* **28**: 501–511.

POTZGER, J. E., and R. C. FRIESNER. 1939. Plant migration in the southern limits of Wisconsin glaciation in Indiana. *Amer. Midl. Nat.* **22**: 351–368.

POTZGER, J. E., and I. T. WILSON. 1941. Post-Pleistocene forest migration as indicated by sediments from three deep inland lakes. *Amer. Midl. Nat.* **25**: 270–289.

POTZGER, J. E., and R. R. RICHARDS. 1942. Forest succession in the Trout Lake, Vilas County, Wisconsin area: A pollen study. *Butler Univ. Bot. Stud.* **5**: 179–189.

POWERS, W. L. 1932. Subsidence and durability of peaty lands. *Agric. Engineering* **13** (3).

RIGG, G. B. 1925. Some Sphagnum bogs of the North Pacific Coast of America. *Ecology* **6**: 260–278.

—— 1940a. Comparisons of the development of some Sphagnum bogs of the Atlantic Coast, the interior, and the Pacific coast. *Amer. Jour. Bot.* **27**: 1–14.

—— 1940b. The development of Sphagnum bogs in North America. *Bot. Rev.* **6**: 666–693.

RIGG, G. B., and C. T. RICHARDSON. 1934. The development of Sphagnum bogs in the San Juan Islands. *Amer. Jour. Bot.* **21**: 617–622.

RIGG, G. B., and C. T. RICHARDSON. 1938. Profiles of some Sphagnum bogs of the Pacific Coast of North America. *Ecology* **19**: 408–434.

ROBERTS, F. H.H., JR. 1935. A Folsom complex: preliminary report on investigations at the Lindenmeier site in northern Colorado. *Smithsonian Misc. Coll.* **96** no. (4).

—— 1936. Additional information on the Folsom complex: report on the second season's investigations at the Lindenmeier site in northern Colorado. *Smithsonian Misc. Coll.* **95** (4).

SAURAMO, MATTI. 1929. The Quaternary geology of Finland. *Bull. de la Commission géol. de Finlande,* **86**.

SAYLES, E. B., and E. ANTEVS. 1941. The Cochise Culture. *Medallion Papers* **29**. Globe, Arizona.

SEARS, P. B. 1930. A record of postglacial climate in northern Ohio. *Ohio Jour. Sci.* **30**: 205–217.

—— 1935. Glacial and postglacial vegetation. *Bot. Rev.* **1**: 37–52.

—— 1941. Postglacial vegetation in the Erie-Ohio area. *Ohio Jour. Sci.* **41**: 225–234.

—— 1942. Xerothermic theory. *Bot. Rev.* **6**: 708–736.

—— 1942a. Postglacial migration of five forest genera. *Amer. Jour. Bot.* **29**: 684–691.

SEARS, P. B., and E. JANSON. 1933. Rate of peat growth in the Erie Basin. *Ecology* **14**: 348–355.

SERNANDER, RUTGER. 1894. Studier öfver den gotländska vegetationens utvecklingshistoria. Doctor's thesis. Upsala.

—— 1908. On the evidence of postglacial changes of climate furnished by the peat-mosses of northern Europe. *Geol. Foren. Forhandl. Bd.* **30**, Haft. 7: 465–473.

—— 1910. Die schwedischen Torfmoore als Zeugen postglazialer Klimaschwankungen. *In* Die Veränderungen des Klimas seit dem Maximum der letzten Eiszelt. *11th Intern. Géol. Congr.*, 195–246. 1911.

SHANTZ, H. L., and R. ZON. 1924. Natural vegetation: Grassland and desert shrub and forest. *Atlas Amer. Agric.*, Washington, D.C. 1936.

SMITH, PRESTON. 1940. Correlations of pollen profiles from glaciated eastern North America. *Amer. Jour. Sci.* **238**: 597–601.

SMITH, W. D. 1933. Physiography of Oregon coast. *Pan-Amer. Geol.* **59**: 33–44; 97–114; 190–206; 241–258.

SPRAGUE, F. L., and H. P. HANSEN. 1946. Forest succession in the McDonald Forest, Willamette Valley, Oregon Northwest. *Science* **20**: 89–99.

STARK, P. 1927. Über die Zugehörigkeit des Kieferpollens in den verscheidenen Horizonten der Bodenseemoore. *Bot. Ges.* **45**: 40–47.

STROM, K. M. 1928. Recent advances in limnology. *Proc. Linn. Soc. London*, 96–110.

SUDWORTH, G. B. 1908. Forest trees of the Pacific slope. U. S. Dept. Agric.

SZAFER, W. 1935. The significance of isopollen lines for the investigation of the geographical distribution of trees in the postglacial period. *Bull. Inter. Acad. Polonaise Sci. Let. Ser. B:* I, 235–239.

TARR, R. S. 1908. Some phenomena of the glacier margins in the Yakutat Bay region, Alaska. *Zeit. F. Gletscherkunde* **3**: 81–110.

THAYER, T. P. 1939. Geology of the Salem Hills and the north Santiam River basin, Oregon. *Oregon Dept. Geol. and Min. Ind., Bull.* **15**: 20–26.

THIENEMANN, A. 1931. Der productionsbegriffin der Biologie. *Arch. hydrobiol.* **22**: 616–622.

THORNBURY, W. D. 1940. Weathered zones and glacial chronology in southern Indiana. *Jour. Geol.* **48**: 449–475.

THORNTHWAITE, C. W. 1931. The climates of North America according to a new classification. *Geog. Rev.* **21**: 633–655.

THWAITES, F. T. 1937. Outline of glacial geology, 1–115. Ann Arbor, Murby.

—— 1943. Pleistocene of part of northeastern Wisconsin. *Geol. Soc. Amer. Bull.* **54**: 87–144.

TRANSEAU, E. N. 1903. On the geographic distribution and ecological relations of the bog plant societies of northern North America. *Bot. Gaz.* **36**: 401–420.

VAN WINKLE, W. 1914. Quality of the surface waters of Oregon. *U. S. Geol. Surv. Water Supply Paper* 363.

VON POST, HAMPUS. 1862. Studier öfver nutidens koprogena jordbildningar, gyttja, dy, torf och mylla. *Kungl. Svenska Vetenskapsakademiens Handlingar*, **4** (1).

VON POST, L. 1930. Problems and working-lines on the post-arctic history of Europe. *Rept. Proc. 5th Intern. Bot. Congr.* 48–54.

—— 1933. Dan svenska skogen efter istiden. *Verdandis Småskrifter* No. 357.

VON SARNTHEIM, R. 1936. Moor-und Seeablagerungen aus den Tiroler Alpen in ihrer waldgeschichtlicher Bedeutung. I. Brennergegund und Fisacktal. *Beih. Bot. Cent.* **55** : 544–631.

VOSS, J. 1931. Preliminary report on the paleoecology of a Wisconsin and an Illinois bog. *Trans. Ill. Acad. Sci.* **24**: 130–137.

—— 1934. Postglacial migration of forests in Illinois, Wisconsin, and Minnesota. *Bot. Gaz.* **96**: 3–43.

WASHBURN, B. 1935. Conquest of Mt. Crillon. *Nat. Geog. Mag.* **67**: 361–300.

WATERS, A. 1939. Resurrected erosion surface in central Washington. *Geol. Soc. Amer., Bull.* **50**: 635–660.

WEAVER, J. E., and F. E. CLEMENTS. 1938. Plant ecology. N. Y., McGraw.

WELCH, P. S. 1935. Limnology. N.Y., McGraw.

WILLIAMS, HOWEL. 1935. Newberry volcano of central Oregon. *Geol. Soc. Amer., Bull.* **46**: 253–305.

—— 1942. Geology of Crater Lake National Park, Oregon. *Carnegie Inst. Wash., Publ.* **540**: 1–157.

—— 1944. Volcanoes of the Three Sisters region, Oregon Cascades. *Univ. Calif. Pub., Bull., Dept. Geol. Sciences* **27** 37–84.

WILLIS, B. 1898. Drift phenomena of Puget Sound. *Geol. Soc. Amer., Bull.* **9**: 111–157.

WILSON, L. R. 1938. The postglacial history of vegetation in northwestern Wisconsin. *Rhodora* **40**: 137–175.

—— 1944. Spores and pollen as microfossils. *Bot. Rev.* **10**: 499–523.

WILSON, L. R., and R. M. KOSANKE. 1940. The microfossils in a pre-Kansan peat deposit near Belle Plain, Iowa. *Torreya* **40**: 1–5.

WILSON, L. R., and R. M. WEBSTER, 1942. Fossil evidence of wider post-Pleistocene range for butternut and hickory in Wisconsin. *Rhodora* **44**: 409–414.

WODEHOUSE, R. P. 1935. Pollen grains. N. Y., McGraw.

WOOLLEY, R. R. 1924. Water powers of the Great Salt Lake basin. *U. S. Geol. Surv. Water Supply Paper* 517.

ZON, see SHANTZ and ZON.

www.ingramcontent.com/pod-product-compliance
Lightning Source LLC
Chambersburg PA
CBHW081337190326
41458CB00018B/6030